Unsupervised and Semi-Supervised Learning

Series Editor

M. Emre Celebi, Computer Science Department, Conway, Arkansas, USA

Springer's Unsupervised and Semi-Supervised Learning book series covers the latest theoretical and practical developments in unsupervised and semi-supervised learning. Titles – including monographs, contributed works, professional books, and textbooks – tackle various issues surrounding the proliferation of massive amounts of unlabeled data in many application domains and how unsupervised learning algorithms can automatically discover interesting and useful patterns in such data. The books discuss how these algorithms have found numerous applications including pattern recognition, market basket analysis, web mining, social network analysis, information retrieval, recommender systems, market research, intrusion detection, and fraud detection. Books also discuss semi-supervised algorithms, which can make use of both labeled and unlabeled data and can be useful in application domains where unlabeled data is abundant, yet it is possible to obtain a small amount of labeled data.

Topics of interest in include:

- Unsupervised/Semi-Supervised Discretization
- Unsupervised/Semi-Supervised Feature Extraction
- Unsupervised/Semi-Supervised Feature Selection
- Association Rule Learning
- Semi-Supervised Classification
- Semi-Supervised Regression
- Unsupervised/Semi-Supervised Clustering
- Unsupervised/Semi-Supervised Anomaly/Novelty/Outlier Detection
- Evaluation of Unsupervised/Semi-Supervised Learning Algorithms
- Applications of Unsupervised/Semi-Supervised Learning

While the series focuses on unsupervised and semi-supervised learning, outstanding contributions in the field of supervised learning will also be considered. The intended audience includes students, researchers, and practitioners.

More information about this series at http://www.springer.com/series/15892

Olfa Nasraoui • Chiheb-Eddine Ben N'Cir
Editors

Clustering Methods for Big Data Analytics

Techniques, Toolboxes and Applications

Springer

Editors
Olfa Nasraoui
Department of Computer Engineering
and Computer Science
University of Louisville
Louisville, KY, USA

Chiheb-Eddine Ben N'Cir
University of Jeddah
Jeddah, KSA

ISSN 2522-848X ISSN 2522-8498 (electronic)
Unsupervised and Semi-Supervised Learning
ISBN 978-3-030-07419-7 ISBN 978-3-319-97864-2 (eBook)
https://doi.org/10.1007/978-3-319-97864-2

Softcover re-print of the Hardcover 1st edition 2019

This Springer imprint is published by the registered company Springer Nature Switzerland AG
The registered company address is: Gewerbestrasse 11, 6330 Cham, Switzerland

Preface

Data has become the lifeblood of today's knowledge-driven economy and society. Big data clustering aims to summarize, segment, and group large volumes and varieties of data that are generated at an accelerated velocity into groups of similar contents. This has become one of the most important techniques in exploratory data analysis. Unfortunately, conventional clustering techniques are becoming more and more unable to process such data due to its high complexity, heterogeneity, large volume, and rapid generation. This raises exciting challenges for researchers to design new scalable and efficient clustering methods and tools which are able to extract valuable information from these tremendous amount of data. The progress in this topic is fast and exciting.

This volume aims to help the reader capture new advances in big data clustering. It provides a systematic understanding of the scope in depth, and rapidly builds an overview of new big data clustering challenges, methods, tools, and applications.

The volume opens with a chapter entitled "Overview of Scalable Partitional Methods for Big Data Clustering." In this chapter, BenHaj Kacem et al. propose an overview of the existing clustering methods with a special emphasis on scalable partitional methods. The authors design a new categorizing model based on the main properties pointed out in the big data partitional clustering methods to ensure scalability when analyzing a large amount of data. Furthermore, a comparative experimental study of most of the existing methods is given over simulated and real large datasets. The authors finally elaborate a guide for researchers and end users who want to decide the best method or framework to use when a task of clustering large scale of data is required.

In the second chapter, "Overview of Efficient Clustering Methods for High-dimensional Big Data Streams," Hassani focuses on analyzing continuous, possibly infinite streams of data, arriving at high velocity such as web traffic data, surveillance data, sensor measurements, and stock trading. The author reviews recent subspace clustering methods of high-dimensional big data streams while discussing approaches that efficiently combine the anytime clustering concept with the stream subspace clustering paradigm. Additionally, novel open-source assessment framework and evaluation measures are presented for subspace stream clustering.

In the chapter entitled "Clustering Blockchain Data," Chawathe gives recent challenges and advances related to clustering blockchain data such as those generated by popular cryptocurrencies like Bitcoin, Ethereum, etc. Analysis of these datasets have diverse applications, such as detecting fraud, illegal transactions, characterizing major services, identifying financial hotspots, characterizing usage and performance characteristics of large peer-to-peer consensus-based systems. The author motivates the study of clustering methods for blockchain data and introduces the key blockchain concepts from a data-centric perspective. He presents different models and methods used for clustering blockchain data and describes the challenges and solutions to the problem of evaluating such methods.

Deep Learning is another interesting challenge, which is discussed in the chapter titled "An Introduction to Deep Clustering" by Gopi et al. The chapter presents a simplified taxonomy of deep clustering methods based mainly on the overall procedural structure or design which helps beginning readers quickly grasp how almost all approaches are designed. This also allows more advanced readers to learn how to design increasingly sophisticated deep clustering pipelines that fit their own machine learning problem-solving aims. Like Deep Learning, deep clustering promises to leave an impact on diverse application domains ranging from computer vision and speech recognition to recommender systems and natural language processing.

A new efficient Spark-based implementation of PSO (particle swarm optimization) clustering is described in a chapter entitled "Spark-Based Design of Clustering Using Particle Swarm Optimization." Moslah et al. take advantage of in-memory operations of Spark to build grouping from large-scale data and accelerate the convergence of the method when approaching the global optimum region. Experiments conducted on real and simulated large data-sets show that their proposed method is scalable and improves the efficiency of existing PSO methods.

The last two chapters describe new applications of big data clustering techniques. In "Data Stream Clustering for Real-Time Anomaly Detection: An Application to Insider Threats," Haider and Gaber investigate a new streaming anomaly detection approach, namely, Ensemble of Random subspace Anomaly detectors In Data Streams (E-RAIDS), for insider threat detection. The investigated approach solves the issues of high velocity of coming data from different sources and high number of false alarms/positives (Fps). Furthermore, in "Effective Tensor-Based Data Clustering Through Sub-tensor Impact Graphs" which completes the volume, Candan et al. investigate tensor-based methods for clustering multimodal data such as web graphs, sensor streams, and social networks. The authors deal with the computational complexity problem of tensor decomposition by partitioning the tensor and then obtain the tensor decomposition leveraging the resulted smaller partitions. They introduce the notion of *sub-tensor impact graphs (SIGs)*, which quantify how the decompositions of these sub-partitions impact each other and the overall tensor decomposition accuracy and present several complementary algorithms that leverage this novel concept to address various key challenges in tensor decomposition.

We hope that the volume will give an overview of the significant progress and the new challenges arising from big data clustering in theses recent years. We also hope that contents will obviously help researchers, practioners, and students in their study and research.

Louisville, KY, USA — Olfa Nasraoui
Manouba, Tunisia — Chiheb-Eddine Ben N'Cir

Contents

Chapter 1
Overview of Scalable Partitional Methods for Big Data Clustering

Mohamed Aymen Ben HajKacem, Chiheb-Eddine Ben N'Cir, and Nadia Essoussi

1.1 Introduction

Clustering, also known as cluster analysis, has become an important technique in machine learning used to discover the natural grouping of the observed data. Often, a clear distinction is made between learning problems that are supervised, also known as classification, and those that are unsupervised, known as clustering [24]. The first deals with only labeled data while the latter deals with only unlabeled data [16]. In many real applications, there is a large supply of unlabeled data but limited labeled data. This fact makes clustering more difficult and more challenging than classification. Consequently, there is a growing interest in a hybrid setting, called semi-supervised learning [11] where the labels of only small portion of the observed data are available.

During the last four decades, many clustering methods were designed based on different approaches such as hierarchical, partitional, probabilistic, and density-based [24]. Among them, Partitional clustering methods have been widely used in several real-life applications given their simplicity and their competitive computational complexity. This category of methods aims to divide the dataset into a number of groups based on the optimization of one, or several objective criteria. The optimized criteria may emphasize a local or a global structure of the data and its optimization is based on an exact or an approximate optimization technique. Despite the competitiveness of the computational complexity of partitional methods compared to other methods, it fails to perform clustering on huge amounts of data

---

M. A. B. HajKacem (✉) · N. Essoussi
LARODEC, Institut Supérieur de Gestion de Tunis, Université de Tunis, Tunis, Tunisia
e-mail: nadia.essoussi@isg.rnu.tn

C.-E. Ben N'Cir (✉)
University of Jeddah, Jeddah, KSA

O. Nasraoui, C.-E. Ben N'Cir (eds.), *Clustering Methods for Big Data Analytics*, Unsupervised and Semi-Supervised Learning,
https://doi.org/10.1007/978-3-319-97864-2_1

[22]. In fact, given the exponential growth and availability of data collected from different sources, analyzing and organizing these data has become an important challenge referred to as *Big data Analytics*. This challenge has been the focus of several types of researches which proposed scalable partitional clustering methods based on different acceleration techniques. The aim of this work is to give a theoretical and empirical overview of scalable big data clustering methods. We propose a new categorization of these methods based on the used methodology to improve scalability. For each defined category, we review the existing methods and we describe the used acceleration technique or framework. Also, an empirical evaluation is given using most representative methods from each of the defined categories using large simulated and real datasets.

The remainder of the chapter is organized as follows. Section 1.2 presents partitional clustering methods. Then, Sect. 1.3 reviews the existing scalable Big data partitional clustering methods. After that, Sect. 1.4 gives an experimental evaluation of Big data partitional clustering methods on different simulated and real large datasets. Finally, Sect. 1.5 presents conclusions and some perspective points.

1.2 Partitional Clustering Methods

More than thousands of clustering methods were proposed in the literature. Commonly used methods are usually classified according to the fundamental concept on which clustering methods are based. This classification leads to defining the following categories [24]: hierarchical, partitional, probabilistic, and density-based. Among them, partitional clustering methods remain the most commonly used because of their simplicity and their competitive computational complexity.

Partitional clustering methods try to organize data into k clusters (where k is an input parameter), by optimizing a certain objective function that captures a local and global structure of grouping. Most of the partitioning methods start with an initial assignment and then use an iterative relocation procedure which moves data points from one cluster to another to optimize the objective function. Examples of these methods are k-means [36], k-modes [26], k-prototypes [25], and fuzzy c-means [7]. The k-means [36] is the most fundamental partitional clustering method which is based on the idea that a center can represent a cluster. After selecting k initial cluster centers, each data point is assigned to a cluster based on a Euclidean distance measure, then k cluster centers are recomputed. This step is repeated until an optimal set of k clusters are obtained based on an objective function. The k-modes [26] uses the simple matching coefficient measure to deal with categorical attributes. The k-prototypes [25] integrates k-means with k-modes methods to partition mixed data. On the other hand, partitional clustering methods often generate partitions where each data point belongs to one and only one cluster. The fuzzy c-means [7] extends this notion to associate each data point a membership function to each cluster.

Despite the efficiency of partitional clustering methods, they do not scale with a huge volume of data [47]. This is explained by the high computational cost to build grouping when dealing with a large amount data. To overcome this weakness, several works were proposed to improve the efficiency of conventional partitional clustering methods. This survey emphasizes on scalable Big data partitional clustering methods, specifically those extending k-means, k-prototypes, and fuzzy c-means methods. We present in the next section an overview of these methods.

1.3 Big Data Partitional Clustering Methods

To deal with large-scale data, several methods were designed in the literature which are based on coupling conventional partitional clustering methods and acceleration techniques. These methods aim to enhance the speed of clustering process by reducing the computational complexity. We propose in this work to classify these methods based on the acceleration technique used to improve the scalability. Four categories are defined, *parallel methods*, *data reduction-based methods*, *centers reduction-based methods*, and *hybrid methods*. Figure 1.1 shows a classification tree of these methods where the depth of the tree represents the progression in time and the width of the tree represents the different categories and subcategories. We detail the following the main characteristics of each category.

1.3.1 Parallel Methods

Parallelization is one of the most used acceleration techniques, which aims to reduce the computational cost of conventional partitional clustering methods. Parallelization is defined as a process where the computation is divided into parallel tasks. Several parallel partitional methods were proposed in the literature. These methods are motivated by the assumption that distance computations between one data point with cluster centers are independent of those between any other data point and cluster centers. Thus, distance computation between different data points and cluster centers can be executed in parallel. The parallelization can be done using different frameworks such as Message Passing Interface (MPI) [46], Graphics Processing Unit (GPU) [38], MapReduce [15], or Spark [50]. In the following, we present an overview of parallel clustering methods which are based on these frameworks.

1.3.1.1 MPI-Based Methods

MPI is a parallel framework designed to process a large amount of data through cluster of machines. It is deployed in master/slave architecture where the master sends tasks to slaves and receives computed results. While slaves receive tasks, process them, and send results to the master. MPI provides set of functions which are used by developers to create parallel applications. These functions aim to communicate and exchange data and messages between machines. For example,

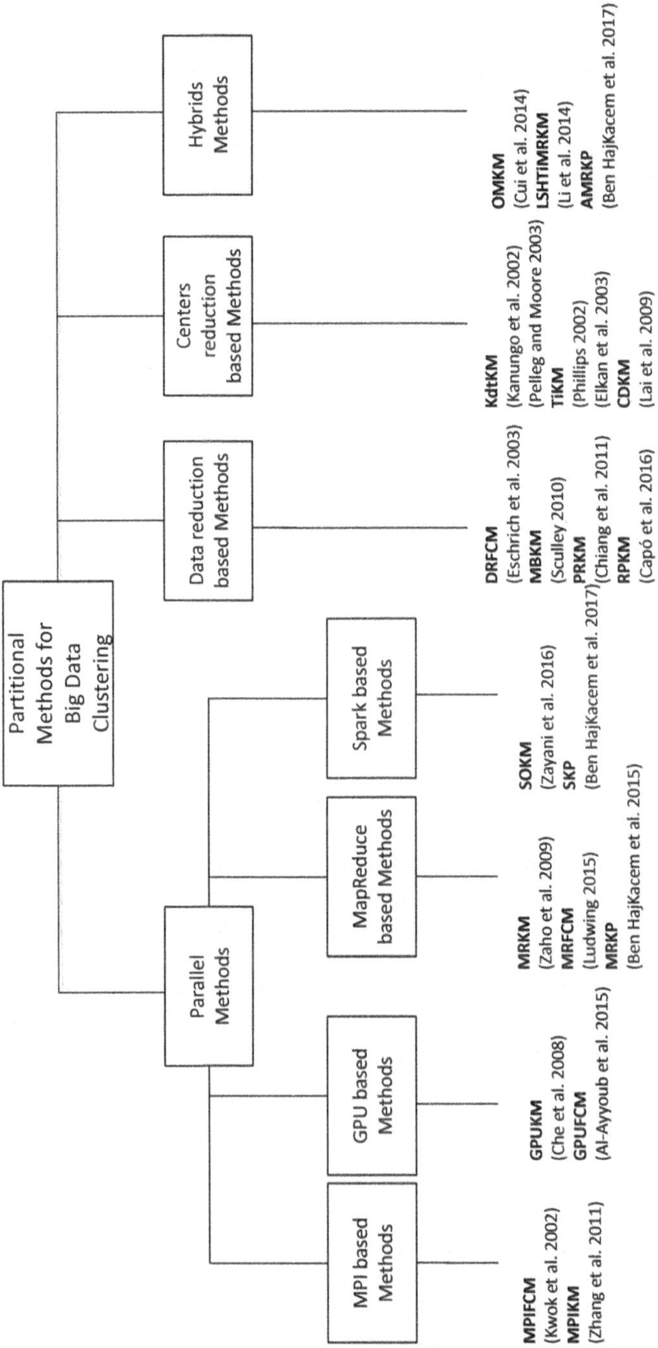

Fig. 1.1 Classification of scalable partitional clustering methods based on the used approach to improve scalability

the function *Broadcast* is used to send the same data or messages to all machines while the function *Barrier* is used to put a barrier and allows the synchronization between machines when running parallel programs. The main advantage of MPI is its master/slave architecture, where the slave machine can become the master of other processes. This can be extremely useful for dynamic resource allocation where the slaves have to process a large amount of data. Another advantage of MPI includes the data preserving, i.e., there is no need to read the same data many times, which can be preserved locally. Hence, MPI is well suited for iterative algorithms [44].

Several methods were proposed to implement clustering process using MPI framework. For example, Zhang et al. [52] proposed an MPI-based k-means (MPIKM). The main steps of this method are as follows: first, the input data is initially distributed among slaves. Then, the master selects k data points as initial centers, and broadcasts them to slaves. Then, each slave assigns corresponding data points to the nearest cluster by computing distances. After that, the master collects all information needed from slaves to update new cluster centers, and broadcasts them to slaves. Finally, this method iterates calling above steps several times until convergence. The MPIKM can reduce the time complexity of k-means from $O(n.k.l)$ to $O(s.k.l)$ where n the number of data points, k the number of clusters, l the number of iterations, and $s \ll n$ the maximal number of data points assigned to each slave node.

On the other hand, Kwok et al. [30] designed a parallelization of fuzzy c-means using MPI framework (MPIFCM). The master first divides the input data into splits and transfers them to slaves. Then, each slave receives the associated split, computes distances, and updates the membership matrix. After that, the master gets all information needed from the slaves to compute the new cluster centers, and sends them to slaves for the next iteration. Finally, the above steps are repeated until convergence. The MPIFCM decreases the time complexity of fuzzy c-means from $O(n.k^2.l)$ to $O(s.k^2.l)$.

Despite the efficiency of MPI framework to process large-scale data, it suffers from limit of the fault intolerance. MPI has no mechanism to handle faults. The failure of one machine in the network can cause the shutdown of the whole network. Hence, practitioners of MPI-based methods have to implement a fault tolerance mechanism within the master/slave architecture to manage faults or failures of machines in the network. The implementation of this mechanism is not a trivial task which explains the restricted use of this framework in real-world applications.

1.3.1.2 GPU-Based Methods

Graphics Processing Unit (GPU) is a specialized hardware designed to accelerate graphical operations such as video and image editing. Compute Unified Device Architecture (CUDA) is a parallel programming framework used to simplify the creation of parallel applications within GPU without delving into the hardware details. GPU has a large number of processing cores as compared to a Central

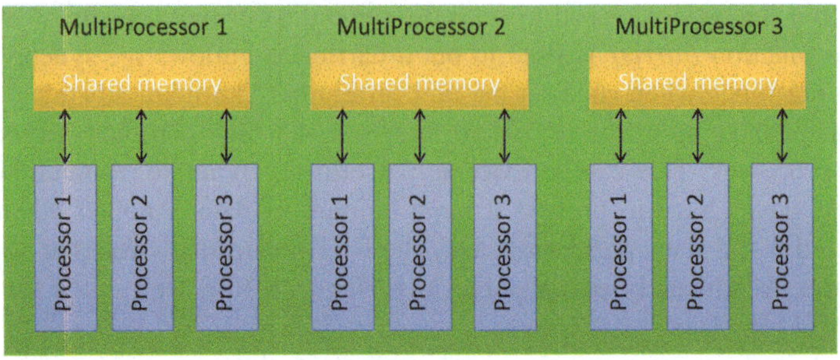

Fig. 1.2 GPU architecture with three multiprocessors and three streaming processors

Processing Unit (CPU). In addition, it provides two levels of parallelization. At the first level, GPU has several multiprocessors (MPs), and at the second level, each multiprocessor has several streaming processors (SPs). Following this configuration, GPU program is broken down into threads which execute on SPs, and these threads are aggregated together to form thread blocks which run on a multiprocessor. Figure 1.2 shows a GPU architecture with three multiprocessors and three streaming processors per block. Each thread within a block can communicate with each other and synchronize with other threads in the same block. Each of these threads has access to fast shared memory.

Several works were proposed to accelerate data clustering using GPU. For example, Che et al. [12] proposed a GPU-based k-means method (GPUKM). The main steps of this method are as follows, first, it uploads initial cluster centers to the shared memory of the GPU and the input dataset is partitioned and uploaded into each multiprocessor. Then, each multiprocessor calculates the distance from each corresponding data point and assigns it to the nearest cluster. After that, it calculates a local cluster center based on a subset of data points. Once all data points are assigned to cluster centers, CPU updates new cluster centers and again will upload them to multiprocessors. Finally, this method iterates calling the above steps several times until convergence. The GPUKM decreases the time complexity of k-means [19] from $O(n.k.l)$ to $O(g.k.l)$ where $g \ll n$ the maximal number of data points assigned to each multiprocessor.

Al-Ayyoub et al. [1] proposed the GPU fuzzy c-means method (GPUFCM). This method first stores initial positions of clusters to the shared memory. Then, it creates initial membership matrix and initial cluster centers from the input data. Then, each multiprocessor computes partial memberships by computing distances. Next, it computes the membership values via summation of partial memberships. After that, this method transfers summed membership values from GPU to CPU in order to compute new cluster centers. Finally, it moves to the next iteration. Similarly to GPUKM, the GPUFCM can reduce the time complexity of fuzzy c-means from $O(n.k^2.l)$ to $O(g.k^2.l)$.

Despite the attested performance of GPU for handling large-scale data, it suffers from the drawback of memory limits. For example, with a maximum of 12GB memory per GPU, it is not able to deal with terabyte data. When the data size exceeds the size of GPU memory, the performance of the GPU-based method decreases significantly. Hence, users have to configure the memory before implementing applications through GPU. Another drawback is the restricted number of software and algorithms that are available for GPUs.

1.3.1.3 MapReduce-Based Methods

MapReduce is a parallel programming framework used to process large-scale data across cluster of machines. It is characterized by its high transparency for programmers to parallelize algorithms in an easy and comfortable way. MapReduce is based on two phases namely *map* and *reduce*. Each phase has $< key/value >$ pairs as input and output. The map phase executes map functions to process in parallel each $< key/value >$ to generate a set of intermediate $< key'/value' >$ pairs. Then, this framework groups all intermediate values associated with the same intermediate key as a list (known as shuffle phase). The reduce phase executes reduce function to merge all intermediate values associated with the same intermediate key. Figure 1.3 illustrates the data flow of MapReduce framework. The inputs and outputs of MapReduce are stored in an associated distributed file system that is accessible from any machine of the used cluster. The implementation of the MapReduce framework is available in Hadoop [48]. Hadoop provides a distributed file system namely Hadoop Distributed File System (HDFS) which stores data on the machines. MapReduce has three major features: simple programming framework, linear scalability, and fault tolerance. These features make MapReduce a useful framework for large-scale data processing.

Several methods were proposed in the literature to fit clustering process through MapReduce. For instance, Zaho et al. [53] proposed a MapReduce-based k-means method (MRKM). Given an input dataset stored in HDFS, this method first divides the input dataset into splits where each split is associated with map function. Then, the map function assigns each data point of the associated split to the nearest cluster by computing distances. The reduce function then updates new cluster centers by calculating the average of data points present in each cluster. These new cluster centers are then written to the HDFS, to be used by the map function for the next iteration. Finally, the entire process is repeated until convergence. The time complexity of MRKM is evaluated by $O(m.k.l)$ where $m \ll n$ the maximal number of data points associated with the map function.

On the other side, Ludwing [35] proposed the parallelization of fuzzy c-means clustering using MapReduce framework (MRFCM). This method is based on two MapReduce jobs. The first MapReduce job calculates cluster center's matrix and the second MapReduce job calculates the distances, to be used to update the membership matrix. The map function of the first MapReduce job receives a chunk of data and a portion of the membership matrix and produces cluster center's sub-matrices. Then, the reduce function of the first MapReduce job merges sub-matrices into cluster center's matrix. The second MapReduce job compared to the first one involves more computations to be executed. During the map function, a chunk of

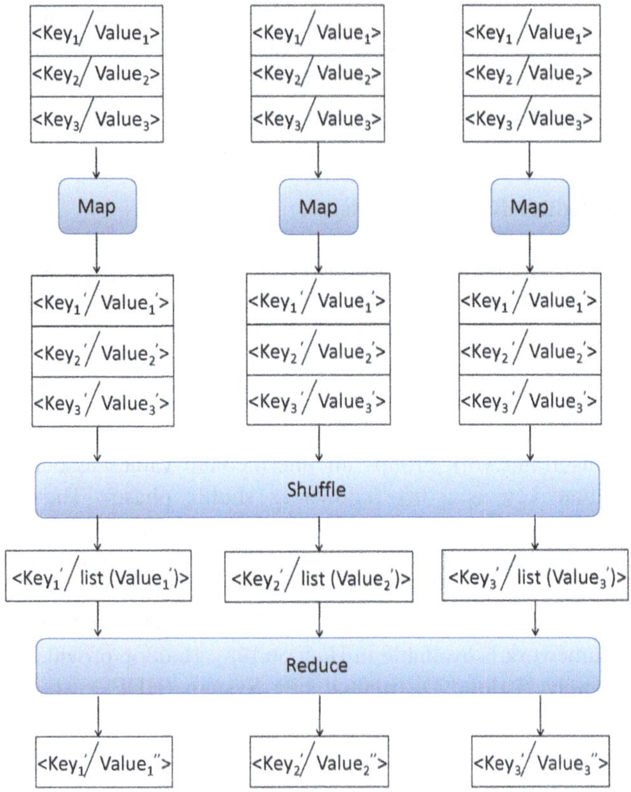

Fig. 1.3 Data flow of MapReduce framework

data is received and distance sub-matrices and membership matrices are computed. Then, the reduce function merges partition sub-matrices. The MRFCM decreases the time complexity of fuzzy c-means from $O(n.k^2.l)$ to $O(m.k^2.l)$.

Ben HajKacem et al. [4] proposed a MapReduce-based k-prototypes (MRKP) for clustering mixed large-scale data. First, MRKP creates a global variable that contains initial centers of the cluster. Then, the map function assigns each data point of the associated split to nearest cluster. Following each assignment, it computes an intermediate center for each cluster in order to optimize the calculation of new centers in reduce phase. This information consists of the sum of the numeric values and the frequencies of categorical values related to each cluster, which are then emitted to the reduce function. After that, the intermediate centers produced from map functions are merged by the reduce function, in order to update new cluster centers. Finally, new centers' values are stored in HDFS and the MRKP moves to the next iteration until convergence. Similarly to MRKM, the time complexity of MRKP is evaluated by $O(m.k.l)$.

Although MapReduce appears to be perfect for clustering large-scale data, it suffers from the inefficiency to run iterative algorithms [18]. The whole dataset must be read and written to HDFS for each iteration of the method. Therefore, many of

Table 1.1 Some operators in Spark

Operators	Meaning
map(**func**)	Iterates over each line in the RDD by calling **func** and returns only one element
flatmap(**func**)	Similar to map but returns a list of elements
mapparition(**func**)	Similar to map but runs separately on each partition (block) of the RDD
filtre(**func**)	Returns elements of the source RDD when **func** returns true
reducebykey(**func**)	Aggregates values with the same key using **func**
distinct()	Returns a new RDD that contains distinct elements of the source RDD

input/output (I/O) disk operations occur during each iteration and this significantly degrades the performance of MapReduce-based method.

1.3.1.4 Spark-Based Methods

Spark is a parallel framework for large-scale data processing designed to solve the MapReduces limitations. It was introduced as part of the Hadoop and it is designed to run with Hadoop, specially by reading data from HDFS. Spark is based on Resilient Distributed Datasets (RDDs), a special type of data structure used to parallelize computations in a transparent way. These parallel structures persist, reuse, and cache the results in memory. Moreover, Spark provides set of in-memory operators, beyond the standard MapReduce, with the aim of processing data more rapidly in distributed environments. Spark is faster up to $100\times$ than MapReduce. Table 1.1 shows some operators of Spark which are used to implement parallel methods.

Several methods were proposed for Big data clustering within Spark framework. For instance, Ben HajKacem et al. [5] proposed a Spark-based k-prototypes clustering method for mixed large-scale data (SKP). The authors exploit in this method the in-memory operations of Spark to alleviate the consumption time of MRKP method. First, they create an RDD object with the input dataset formed by c chunks. The map function (mappartition) picks a chunk of dataset, executes the k-prototypes algorithm on that chunk, and emits the intermediate cluster centers as output. Then, the reduce function (reducebykey) takes set of intermediate centers, executes the k-prototypes algorithm again on them, and returns the final centers as output. For each map phase, there are m data points that must be processed using k-prototypes. Hence, map phase takes $O(m.k.l)$ time. In the reduce phase, the k-prototypes must be executed on set of intermediate centers which has $k.c$ items. Hence, the reduce phase needs $O(c.k^2.l)$ time. Given $c \ll m$, the overall time complexity of SKP is evaluated by $O(m.k.l)$.

Zayani et al. [51] proposed a parallelization of overlapping k-means method using Spark framework (SOKM). This method can perform parallel clustering processes leading to non-disjoint partitioning of data. The main steps of this method are as follows: first, they create an RDD object with input data formed by chunks.

Then, the map function (map) performs a local data assignment for each chunk in parallel in order to build the local membership matrix for set of data points within this chunk. Once local membership matrix is computed for each chunk, the second map function (map) takes these intermediate results and updates the local cluster centers. After that, the reduce function (reducebykey) computes global cluster centers based on intermediate cluster centers computed on each chunk. Finally, based on the evaluation of the global objective criterion, the proposed method reiterates all above-described steps while convergence is not reached. The time complexity of SOKM is evaluated by $O(m.k^2.l)$.

1.3.2 Data Reduction-Based Methods

This category of methods tries to reduce the number of data points when building clusters in order to accelerate the clustering process. Sculley [43] introduced a MinBatch k-means method (MBKM). Its main idea is to use small random batches of data points of a fixed size which can be stored in memory. The motivation behind this method is that random batches tend to have lower stochastic noise than individual data points. In each iteration, a new random sample from data is generated and used to build cluster centers. The cluster centers are then updated using a convex combination of values of cluster centers and data points by applying a dynamic rate. This dynamic rate is defined for each cluster and is evaluated by the inverse of the number of assigned data points. The MBKM decreases the time complexity of k-means from $O(n.k.l)$ to $O(b.k.l)$ where $b \ll n$ the batch size.

On the other hand, Capo et al. [9] proposed an alternative to the k-means for processing large-scale data called Recursive Partition k-means (RPKM). This method considers a sequence of partitions of the dataset, where the partition at iteration i is thinner than the partition at iteration $i - 1$. The idea behind this method is to approximate the k-means for the entire dataset by recursively applying a weighted version of k-means over a small number of subsets of data. The main steps of RPKM can be described as follows: first, the data is partitioned into a number of subsets where each subset is characterized by a representative and its corresponding weight. Second, a weighted version of k-means is applied over the set of representatives. From one iteration to the next, a more refined partition is built and the process is repeated using the obtained optimal cluster centers as initialization. The above steps are repeated until a stopping criterion is detected. The RPKM can reduce the time complexity of k-means from $O(n.k.l)$ to $O(p.k.l)$ where $p \ll n$ the partition size.

PRKM method proposed a Pattern Reduction algorithm for reducing the computation time of k-means [13]. Initially, the PRKM works exactly as k-means. Then, it continues to check whether it is time to start the pattern reduction algorithm. If the time to start is reached, the pattern reduction algorithm is applied. The pattern reduction algorithm is based on compressing and removing at each iteration the data points that are unlikely to change their membership thereafter. It can be divided into two procedures: the first is Pattern Compression and Removal (PCR), and the second is Pattern Assignment and Mean Update (PAMU). PCR requires a removal bound parameter which denotes the percentage of data points that are allowed to be compressed and removed. If the removal bound has not been reached, PCR first checks which data points in each cluster are near the mean and thus can be removed. We note that means is defined as the average of distances between data points and cluster centers. Then, PCR compresses and removes these data points by selecting one of data points to be removed as the representative data point and setting its value to the average of all data points removed. After that, PAMU reassigns each data point to the cluster to which it belongs first and then computes the new cluster centers. The PRKM can reduce the time complexity of k-means from $O(n.k.l)$ to $O(n.k)$.

Another method introduced a modified version of fuzzy c-means algorithm that uses Data Reduction algorithm to cluster large-scale data (DRFCM) [20]. The main idea of DRFCM is to reduce the data size before performing clustering. This method has two main phases, data reduction and data clustering. The data reduction phase consists of two steps which are precision reduction (an optional step) and aggregation. The aggregation step consists of combining identical data points together in the same weighted data points where the weights correspond to the number of aggregated data points. The clustering phase is devoted to apply the fuzzy c-means algorithm on the new weighted data points. To improve the running time of the aggregation step, the authors introduced a hashing for the aggregation step. Moreover, they optimize the calculation of cluster centers and membership matrix. The DRFCM decreases the time complexity of fuzzy c-means from $O(n.k^2.l)$ to $O(w.k^2.l)$ where $w \ll n$ the number of weighted data points generated by data reduction algorithm.

1.3.3 Centers Reduction-Based Methods

This category of methods aims to reduce the number of comparisons when looking for the nearest cluster centers which is the most time-consuming phase. Pelleg and Moore [39] and Kanungo et al. [28] proposed to accelerate k-means using kd-tree structure (KdtKM). A kd-tree is defined as a binary tree that partitions the space of the data and it is built by separating hyperplanes (typically axis-aligned) [42]. We note that k in kd-trees and in k-means are different. In k-means, it denotes the number of clusters while in kd-trees, denotes the dimension of the input data.

Pelleg and Moore [39] proposed to represent data as a kd-tree and used a pruning strategy to reduce the redundant comparisons between cluster centers and data points. This method initially creates the kd-tree from the input dataset. Then, it performs at each iteration a search through the tree for searching regions of the tree which are *owned* by a single center. The search begins with all k cluster centers at the root of the kd-tree. Then, it recursively goes through the depth of the tree while checking at each node that only one center *dominates* all the other centers. If so, it eliminates the dominated centers. In the case that only exists one dominated center, then the search stops and all data points below that node in the kd-tree are assigned to that center. However, when multiple centers remain, it recursively continues its search on child nodes. If it reaches a leaf, it performs distance computations between the remaining cluster centers and the data points at the leaf node. This method finds nearest cluster to each data point in $O(\log k)$ time. However, it requires $O(k.\log k)$ time in each iteration for building the kd-tree. Hence, it decreases the time complexity of k-means from $O(n.k.l)$ to $O(n.\log k.l+k.\log k)$. Although KdtKM method performs very well when dealing with large-scale data, it is not suitable for high-dimension data because of its exponential complexity regarding the number of dimensions of data. In addition, it requires an extra memory on the order of the input dataset for storing the kd-tree structure.

Other techniques are used to accelerate partitioning methods such as Triangle inequality-based k-means (TiKM) methods [17, 19, 23, 41]. For example, Phillips [41] used the triangle inequality to reduce the number of comparisons between cluster centers and data points. This method is motivated by the fact that k-means requires computing all distances between each of the cluster centers and data points. Many of these distance computations are redundant, because data points usually stay in the same cluster after first few iterations. This method uses the following triangle inequality to prove that if cluster center c_1 is close to data point x, and some other cluster center c_2 is far away from another cluster center c_1, then c_1 must be closer than c_2 to x.

Theorem 1.1 *Let x a data point and let c_1 and c_2 cluster centers (described in Fig. 1.4). If we know that $d(c_1,c_2) \geq 2* d(x,c_1)$, then $d(x,c_1) \leq d(x,c_2)$ without having to calculate $d(x,c_2)$.*

In each iteration, the TiKM method uses the triangle inequality when looking for nearest cluster centers. Also, it requires computing distances between cluster centers. Hence, this method decreases the time complexity of k-means from $O(n.k.l)$ to $O((n.\gamma+k^2).l)$ where γ the average number of comparisons between data points and clusters selected at each iteration.

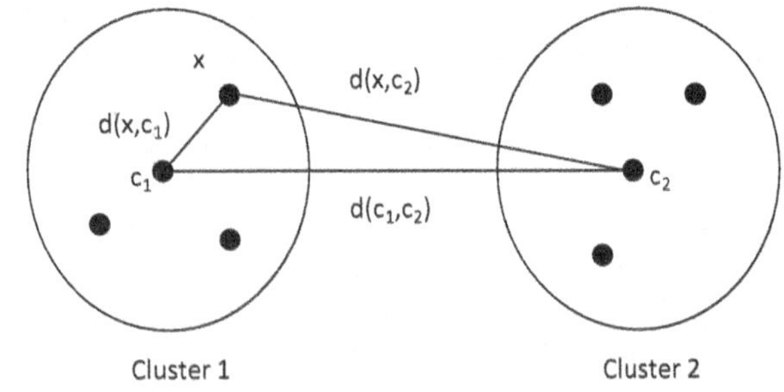

Fig. 1.4 The application of triangle inequality technique between data point x and the cluster centers c_1 and c_2

On the other hand, a Center Displacement k-means method (CDKM) is proposed to improve the efficiency of k-means [31]. This method first classifies cluster centers into static and active groups. Then, it uses the information of center displacements to reject impossible candidates when locking to the nearest cluster center. The CDKM decreases the time complexity of k-means from $O(n.k.l)$ to $O(n.k)$. We note that the time complexity of CDKM grows linearly with the data dimension in contrast to KdtKM methods which have the exponential dependence on the value of dimension. Moreover, CDKM obtains the same set of cluster centers as that produced by the conventional k-means.

1.3.4 Hybrids Methods

To deal with large-scale data, hybrids methods combine several acceleration techniques. The acceleration techniques are combined to win maximal efficiency when designing a clustering process for analyzing a large amount of data. Example of hybrids methods is the OMRKM proposed by Cui et al. [14]. This method proposed an optimized implementation of MapReduce-based k-means method using sampling. OMRKM consists of three MapReduce jobs namely: *Data sampling*, *Data clustering*, and *Data assignment*. The first MapReduce job is devoted generating a subset from input data using probability sampling. The second MapReduce job is concerned with clustering of the subset in order to obtain cluster centers. Once the cluster centers are computed, the third MapReduce job is executed to generate the partition matrix of input data by assigning each data point to the nearest cluster center. The OMRKM decreases the time complexity of k-means from $O(n.k.l)$ to $O(m{+}r.k.l)$ where $r \ll n$ the sample size.

On the other hand, Li et al. [33] proposed LSHTiMRKM method which is based on MapReduce framework, locality sensitive hashing [27] (LSH), and triangle inequality to improve the efficiency of k-means. This method consists of two MapReduce jobs namely: *Data skeleton* and *Data clustering*. The first job, data skeleton uses the LSH to map similar data points into buckets. Each bucket is represented by a weighted data point. Hence, the LSH technique is used to reduce the number of data points when building cluster centers. The second job, data clustering proposes an efficient implementation of scalable k-means [2]. This implementation is based on a pruning strategy to accelerate the iteration process by reducing unnecessary comparisons between cluster centers and data points. The pruning strategy begins by looking for the nearest cluster centers using triangle inequality. Then, it uses the local property of LSH to reduce distance computations between data points and cluster centers. Only centers in the bucket are evaluated. The LSHTiMRKM decreases the time complexity of k-means from $O(n.k.l)$ to $O(m{+}t.\gamma.l)$ where $t \ll n$ the number of buckets and $\gamma \ll k$ the average number of comparisons between data points and clusters selected at each iteration.

Ben HajKacem et al. [6] proposed an accelerated MapReduce-based k-prototypes for clustering mixed large-scale data (AMRKP). This method is based on a pruning strategy (PS) to reduce redundant distance computations between cluster centers and data points using triangle inequality technique. AMRKP introduces a KPPS algorithm which is based on applying a pruning strategy to k-prototypes. Initially, the KPPS works exactly as k-prototypes. Then, it continues to check whether it is time to start the pruning strategy. If the time to start is reached, the pruning strategy is applied. The pruning strategy requires at each iteration computing distances between centers and sorting them. Then, it evaluates the triangle inequality between data point and the centers in increasing order of distance to the assigned center of the previous iteration. If the pruning strategy reaches a center that does not satisfy the triangle inequality property, it can skip all the remaining centers and continue on to the next data point. The pruning strategy requires setting a pruning bound parameter (α) to denote the $\alpha\%$ subset of cluster centers that are considered when evaluating triangle inequality property. After that, this method distributes the KPPS algorithm within MapReduce framework. For this purpose, it first splits the input dataset into p chunks. Then, the map function picks a chunk of data, executes the KPPS algorithm on that chunk, and emits the intermediate cluster centers as the output. Then, the reduce function takes the set of intermediate centers, executes again the KPPS algorithm on them, and returns the final centers as output. The time complexity of KPPS algorithm is bounded between $O((n.\alpha.k{+}k^3).l)$ and $O((n.k{+}k^3).l)$ where α the pruning bound. The KPPS algorithm is applied twice: in the map phase and the reduce phases. In the map phase, each chunk involves running the KPPS algorithm on that chunk. This phase is evaluated by $O(m.\alpha.k{+}k^3).l)$ time. In the reduce phase, the KPPS algorithm must be executed on the set of intermediate centers which has $k.p$ data points. Hence, the reduce phase needs $O((m.\alpha.k{+}k^3).l)$ time. Given that $\frac{k.n}{p} \ll m$, the overall time complexity of the AMRKP is evaluated by $O((m.\alpha.k{+}k^3).l{+}\left(\frac{k.n}{p}.\alpha.k + k^3\right).l) \cong O((m.\alpha.k{+}k^3).l)$.

1.3.5 Summary of Scalable Partitional Clustering Methods for Big Data Clustering

This section gives an overview of the main characteristics of scalable partitional clustering methods presented in a comparative way. Table 1.2 summarizes these main characteristics. Our study is based on the following features of the methods: (1) type of data supported by each method, (2) the final results after acceleration regarding the conventional method, exact or approximate (3) time complexity, (4) space complexity, and (5) type of the used acceleration technique.

Parallel partitional clustering methods are divided into four categories MPI-based, GPU-based, MapReduce-based, and Spark-based methods. Before fitting clustering through parallel framework, it is important to consider some points. GPU suffers from the memory limitation. When the data size exceeds the size of the GPU memory, the performance decreases significantly the GPU-based method. For example, with a maximum of 12GB memory per GPU, it is not suitable to deal with terabyte data. Then, MPI has no mechanism to handle faults. The failure of one machine in the network can cause the shutdown of the whole network. Hence,

Table 1.2 The main characteristics of scalable partitional clustering methods for Big data clustering

Category	Method	References	Type of data	Results after acceleration	Time complexity	Space complexity	Acceleration technique
Parallel	MPIKM	Zhang et al. [52]	Numeric	Exact	$O(s.k.l)$	$O(s)$	MPI framework
	MPIFCM	Kwok et al. [30]	Numeric	Exact	$O(s.k^2.l)$.	$O(s)$	MPI framework
	GPUKM	Che et al. [12]	Numeric	Exact	$O(g.k.l)$	$O(g)$	GPU framework
	GPUFCM	Al-Ayoub et al. [1]	Numeric	Exact	$O(g.k.l)$	$O(g)$	GPU framework
	MRKM	Zaho et al. [53]	Numeric	Exact	$O(m.k.l)$	$O(m)$	MapReduce framework
	MRFCM	Ludwing [35]	Numeric	Exact	$O(m.k^2.l)$	$O(m)$	MapReduce framework
	MRKP	Ben HajKacem et al. [4]	Mixed	Exact	$O(m.k.l)$	$O(m)$	MapReduce framework
	SKP	Ben HajKacem et al. [5]	Mixed	Approximate	$O(m.k.l)$	$O(m)$	Spark framework
	SOKM	Zayani et al. [51]	Numeric	Exact	$O(m.k.l)$	$O(m)$	Spark framework
Data reduction based	MBKM	Scully [43]	Numeric	Approximate	$O(b.k.l)$	$O(n+b)$	Min batch
	RPKM	Capo et al. [9]	Numeric	Approximate	$O(p.k.l)$	$O(n+p)$	Recursive partition
	PRKM	Chiang et al. [13]	Numeric	Approximate	$O(n.k)$	$O(n)$	Pattern reduction
	DRFCM	Eschirh et al. [20]	Numeric	Exact	$O(w.k.l)$	$O(w)$	Data reduction
Centers reduction based	KdtKM	Pelleg and More [40]	Hard	Numeric	$O(n.\log k.l+k.\log k)$	$O(n+k.\log k)$	Kd-tree
	KdtKM	Kanungo et al. [28]	Numeric	Exact	$O((n.\gamma+k^2).l)$	$O(n+k^2)$	Kd-tree
	TiKM	Philips [41]	Numeric	Exact	$O((n.\gamma+k^2).l)$	$O(n+k^2)$	Triangle inequality
	TiKM	Elkan [19]	Numeric	Exact	$O((n.\gamma+k^2).l)$	$O(n+k^2)$	Triangle inequality
	CDKM	Lai et al. [31]	Numeric	Exact	$O(n.k)$	$O(n+k^2)$	Displacement center
Hybrids	OMRKM	Cui et al. [14]	Numeric	Approximate	$O(m.k+r.k.l)$	$O(n+r)$	MapReduce + Sampling
	LSHTiMRKM	Li et al. [33]	Numeric	Approximate	$O(m+t.\gamma.l)$	$O(n+t+k^2)$	MapReduce + LSH
	AMRKP	Ben HajKacem et al. [6]	Mixed	Approximate	$O(m.\alpha\%.k+k^3).l)$	$O(m+k^2)$	MapReduce + Triangle inequality

Notes: n: number of data points, k: number of clusters, l: number of iterations, s: number of data points assigned to salve, g: number of data points assigned to multiprocessor, m: number of data points assigned to map, b: batch size, p: partition size, w: number of weighted examples, γ: average number of comparisons, r: sample size, t: bucket size, $\alpha\%$: pruning bound practitioners have to implement some kind of fault tolerance mechanism within the program to overcome faults. The MapReduce framework looks better than MPI since it is characterized by simple programming framework, linear scalability, and fault tolerance. However, it is unsuitable to run iterative algorithms since at each iteration, the whole dataset must be read and written to disks and this results in high (I/O) operations. This significantly degrades the performance of MapReduce-based method. Finally, Spark framework is an alternative to MapReduce which is designed to overcome the disk I/O limitations and improve the performance of MapReduce framework. A recent survey has presented the different frameworks for Big data analytics and provides the advantages and drawbacks of each of these frameworks based on various metrics such as scalability, data I/O rate, fault tolerance, and iterative task support [37, 45].

Although all the described Big data partitional clustering methods offer for users an efficient analysis for large-scale data, some parameters need to be estimated before performing the learning. All the described methods require to configure the number of clusters in prior which is not a trivial task in real-life applications where the number of expected clusters is usually unknown. As a solution, one could use different model heuristics for determining the optimal number [34, 40]. For example, the user can test different clustering with an increased number of clusters and then, take the clustering having the best balance between the minimization of the objective function and the number of clusters. Furthermore, all the described Big data partitional methods need to initialize the clusters' centers. However, high-quality initialized centers are important for both accuracy and efficiency of conventional clustering methods. To overcome this problem, users can adopt random sampling method to obtain cluster centers or using initialization techniques which exploit the fact that a good clustering is relatively spread out [8, 10, 32]. Using center initialization techniques, the result of the presented methods always converges to a local optimum of the objective criterion, rather than the global optimum. To deal with this issue, users can combine conventional methods with heuristic techniques to prevent clustering results from falling into local optimum [3, 21, 29].

1.4 Empirical Evaluation of Partitional Clustering Methods for Large-Scale Data

We evaluate in this section the performance of the scalable partitional clustering methods. We select from each category of scalable Big data partitional clustering methods at least one representative method, such as MRKM [53] (parallel), SKP [5] (parallel), MBKM [43] (data reduction based), TiKM [41] (centers reduction based), and AMRKP [6] (hybrids). We have implemented all these methods with the Java version 8 programming language. For MBKM and TiKM, we used a single machine with 1-core 3.4 GHz $i5$ CPU and 4GB of memory while for MRKM, SKP, and AMRKP, we used a cluster of four machines where each machine has 1-core

2.3 GHz CPU and 1GB of memory. Concerning the implementation of MapReduce framework, we used Apache Hadoop version 1.6.0 for MRKM and AMRKP and we used Apache Spark version 1.6.2 for SKP.

The experiments are performed on simulated and real datasets. For simulated datasets, we generate two series of datasets with Gaussian distribution. The mean of generated data points of the Gaussian distribution is 350 and the sigma is 100. The datasets range from 5 to 10 million data points. Each data point is described using ten attributes. In order to simplify the names of simulated datasets, we will use the notations Sim5M and Sim10M to denote a generated dataset containing 5 and 10 million data points, respectively.

Concerning real datasets, we use the KDD Cup dataset (KDD), which consists of normal and attack connections simulated in a military network environment. The KDD dataset contains about 5 million connections. Each connection is described using 33 numeric attributes. The clustering process for these dataset detects the type of attacks among all connections. This dataset was obtained from UCI machine learning repository.[1] The second real dataset is the Household dataset (House), which contains the results of measurements of electric power consumption in household. The House dataset contains 1 million data points. Each data point is described using seven numeric attributes. The clustering process for these data identifies the types of electric consumption in household. This dataset was obtained from UCI machine learning repository.[2] Statistics of simulated and real datasets are summarized in Table 1.3.

To simplify the discussion of the empirical results in Table 1.4, we will use the following conventions, let ψ denotes one of the scalable partitional clustering methods, let T denote the running time, and let S denote the quality of the clustering results. The enhancement of the running time of scalable partitional clustering method (T_ψ) with respect to the running time of k-means (T_{KM}) in percentage is defined by:

$$\Delta_T = \frac{T_\psi - T_{\mathrm{KM}}}{T_{\mathrm{KM}}} * 100\% \tag{1.1}$$

For example, the enhancement of the running time of MRKM (T_{MRKM}) with respect to the running time of k-means is defined by:

Table 1.3 Summary of datasets

Dataset	Data points	Attributes	Domain
Sim5M	5,000,000	10	Simulated
Sim10M	10,000,000	10	Simulated
KDD	4,898,431	33	Intrusion detection
House	1,000,000	7	Electricity

[1] https://archive.ics.uci.edu/ml/datasets/KDD+Cup+1998+Data.

[2] https://archive.ics.uci.edu/ml/datasets/Individual+household+electric+power+consumption.

$$\Delta_T = \frac{T_{\text{MRKM}} - T_{\text{KM}}}{T_{\text{KM}}} * 100\% \tag{1.2}$$

It is important to note that as defined in Eq. 1.2, a more negative value of Δ_T implies a greater enhancement. The enhancement of the quality of scalable partitional clustering (S_ψ) with respect to the quality of k-means (S_{KM}) in percentage is defined by:

$$\Delta_S = \frac{S_\psi - S_{\text{KM}}}{S_{\text{KM}}} * 100\% \tag{1.3}$$

To evaluate the clustering quality, we use the Sum Squared Error (SSE) [49], which aims to measure the squared errors between each data point and the cluster center to which the data point belongs. SSE can be defined by:

$$\text{SSE} = \sum_{i=1}^{n} \sum_{j=1}^{k} d(c_j, x_i), \tag{1.4}$$

where x_i is the data point and c_j the cluster center.

Table 1.4 reports the obtained results of scalable partitional clustering methods compared to k-means on simulated and real datasets. We note that we test for each dataset three different number of clusters 10, 50, and 100 and we fix the number of iterations to 20. For other parameters, we consider the following: the batch size b to 10,000 for MBKM and the pruning bound α to 10 for AMRKP.

The analysis of the empirical results firstly shows that hybrids methods are significantly faster than all other methods since they use simultaneously several acceleration techniques to improve the efficiency of conventional k-means method. For instance, we can observe that TiKM reduces the running time by 35.33% while AMRKP reduces the running time by 85.60%. Hence, we can conclude that the combination of acceleration techniques is a good solution when dealing with large-scale data. The second conclusion concerns the parallel partitional clustering methods. The results show that SKP method is faster than MRKM method. For example, MRKM can reduce the running time of k-means by 75.15%. However, SKP can reduce the running time of k-means by 96.14%. This fact shows the benefits of Spark to execute clustering process in memory and reduce the I/O operations from disks. Hence, we can conclude that Spark framework is more suitable to cluster large-scale data than MapReduce framework. Third, empirical results show that the number of clusters has an impact on the performance of clustering especially the case of triangle inequality-based methods like TiKM and AMRKP. For example, TiKM reduces the running time of k-means by 53.66% when $k = 10$ while by 59.92% when $k = 100$. From this observation, we can mention that TiKM method performs well when dealing with large number of clusters. The fourth conclusion concerns the quality of the obtained results. The results show that MRKM and TiKM methods produce the same SSE values compared

Table 1.4 Empirical results on simulated and real datasets

		MRKM-KM		SKP-KM		MBKM-KM		TiKM-KM		AMRKP-KM	
Dataset	k	Δ_T	Δ_S	Δ_T	Δ_S	Δ_T	Δ_S	Δ_T	Δ_S	Δ_T	Δ_S
Sim5M	10	−75.15	0.00	−96.49	21.03	−98.38	−0.03	−35.33	0.00	−82.60	58.19
	50	−77.14	0.00	−96.25	28.36	−98.45	0.17	−29.33	0.00	−84.83	63.55
	100	−76.18	0.00	−96.77	154.84	−98.63	4.28	−28.64	0.00	−84.69	−65.70
Sim10M	10	−71.25	0.00	−96.52	20.71	−92.12	−0.03	−39.24	0.00	−84.45	58.40
	50	−71.25	0.00	−96.82	28.06	−98.11	1.27	−29.08	0.00	−83.76	63.56
	100	−71.25	0.00	−97.04	27.32	−98.87	−2.25	−36.17	0.00	−83.21	65.68
KDD	10	−75.12	0.00	−96.62	180.70	−96.83	45.27	−53.66	0.00	−85.82	98.50
	50	−75.12	0.00	−96.19	134.30	−98.12	73.04	−56.66	0.00	−87.12	98.14
	100	−75.12	0.00	−96.40	142.61	−98.20	73.88	−59.92	0.00	−88.55	98.50
House	10	−75.56	0.00	−92.99	25.87	−93.70	4.33	−59.32	0.00	−85.50	6.20
	50	−75.56	0.00	−93.48	25.74	−94.05	−2.47	−61.35	0.00	−83.96	7.58
	100	−75.56	0.00	−94.74	12.36	−94.80	0.88	−62.55	0.00	−83.14	6.74

to k-means method. Therefore, we can deduce that MapReduce framework and triangle inequality reduce the running time of conventional k-means method without affecting clustering results. However, MBKM method does not always converge to the same local solution because of the relatively small batch size, leading to significant variations in the SSE values.

As a summary, the empirical study allows us to draw the following conclusions:

- The hybrids methods are faster than all other methods since they use simultaneously several acceleration techniques to improve the efficiency of conventional methods. Therefore, the combination of acceleration techniques is a good solution when dealing with large-scale data.
- Spark framework is designed to overcome disk I/O limitations and supports iterative algorithms. Hence, it is more suitable than MapReduce framework for processing large-scale data.
- The triangle inequality-based methods perform well when the number of clusters is large. However, they require additional memory to store the distances between cluster centers.
- MapReduce and triangle inequality techniques can improve the efficiency of conventional clustering without affecting the final clustering results. While sampling-based methods like MBKM do not always converge to the same local solution since they use a subsample instead of the entire dataset.

1.5 Conclusion

We focused in this chapter the area of Big data clustering, for which we give a classification of the existing methods based on the used acceleration techniques. Our study is essentially based on partitional clustering methods. For that, we review the existing scalable Big data partitional methods in the literature, classified into four main categories: parallel methods, data reduction-based methods, centers reduction-based methods, and hybrids methods. We also provide theoretical and experimental comparisons of the existing Big data partitional methods.

At the end of this overview, we claim that Big data clustering has a growing interest in machine learning research since many real-world applications require a scalable partitioning from a large amount of data. Many current challenges in Big data clustering area motivate researchers to propose more effective and efficient learning process. For example, recent works are interested in the identification of groups from streaming large-scale data, or building clusters from uncertain large-scale data. Other works are interested in the identification of clustering from data having multiple representations. All these challenges within Big data clustering open exciting directions for future researchers.

References

1. M. Al-Ayyoub, A.M. Abu-Dalo, Y. Jararweh, M. Jarrah, M. Al Sa'd, A GPU-based implementations of the fuzzy C-means algorithms for medical image segmentation. J. Supercond. **71**(8), 3149–3162 (2015)
2. B. Bahmani, B. Moseley, A. Vattani, R. Kumar, S. Vassilvitskii, Scalable k-means++. Proc. VLDB Endow. **5**(7), 622–633 (2012)
3. S. Bandyopadhyay, U. Maulik, An evolutionary technique based on K-means algorithm for optimal clustering in RN. Inform. Sci. **146**(1), 221–237 (2002)
4. M.A. Ben HajKacem, C.E. Ben N'cir, N. Essoussi, MapReduce-based k-prototypes clustering method for big data, in *Proceedings of Data Science and Advanced Analytics*, pp. 1–7 (2015)
5. M.A. Ben HajKacem, C.E. Ben N'cir, N. Essoussi, KP-S: a spark-based design of the K-prototypes clustering for big data, in *Proceedings of ACS/IEEE International Conference on Computer Systems and Applications*, pp. 1–7 (2017)
6. M.A. Ben HajKacem, C.E. Ben N'cir, N. Essoussi, One-pass MapReduce-based clustering method for mixed large scale data. J. Intell. Inf. Syst. 1–18 (2017)
7. J.C. Bezdek, R. Ehrlich, W. Full, FCM: the fuzzy c-means clustering algorithm. Comput. Geosci. **10**(2–3), 191–203 (1984)
8. P.S. Bradley, U.M. Fayyad. Refining initial points for K-means clustering, in *Proceeding ICML '98 Proceedings of the Fifteenth International Conference on Machine Learning*, vol. 98, pp. 91–99 (1998)
9. M. Capó, A. Pérez, J.A. Lozano, An efficient approximation to the k-means clustering for massive data. Knowl.-Based Syst. **117**, 56–69 (2017)
10. M.E. Celebi, H.A. Kingravi, P.A. Vela, A comparative study of efficient initialization methods for the k-means clustering algorithm. Expert Syst. Appl. **40**(1), 200–210 (2013)
11. O. Chapelle, B. Scholkopf, A. Zien, Semi-supervised learning (Chapelle, O. et al., Eds.; 2006)[Book reviews]. IEEE Trans. Neural Netw. **20**(3), 542–542 (2009)
12. S. Che, M. Boyer, J. Meng, D. Tarjan, J.W. Sheaffer, K. Skadron, A performance study of general-purpose applications on graphics processors using CUDA. J. Parallel Distrib. Comput. **68**(10), 1370–1380 (2008)
13. M.C. Chiang, C.W. Tsai, C.S. Yang, A time-efficient pattern reduction algorithm for k-means clustering. Inform. Sci. **181**(4), 716–731 (2011)
14. X. Cui, P. Zhu, X. Yang, K. Li, C. Ji, Optimized big data K-means clustering using MapReduce. J. Supercomput. **70**(3), 1249–1259 (2014)
15. J. Dean, S. Ghemawat, MapReduce: simplified data processing on large clusters. Commun. ACM **51**(1), 107–113 (2008)
16. R.O. Duda, P.E. Hart, D.G. Stork, *Pattern Classification*. Wiley, Hoboken (2012)
17. J. Drake, G. Hamerly, Accelerated k-means with adaptive distance bounds, in *5th NIPS Workshop on Optimization for Machine Learning*, pp. 42–53 (2012)
18. J. Ekanayake, H. Li, B. Zhang, T. Gunarathne, S.H. Bae, J. Qiu, G. Fox, Twister: a runtime for iterative MapReduce, in *Proceedings of the 19th ACM International Symposium on High Performance Distributed Computing*, pp. 810–818 (ACM, 2010)
19. C. Elkan, Using the triangle inequality to accelerate k-means, in *Proceeding ICML'03 Proceedings of the Twentieth International Conference on International Conference on Machine Learning*, vol. 1(3) (2003), pp. 147–153
20. S. Eschrich, J. Ke, L.O. Hall, D.B. Goldgof, Fast accurate fuzzy clustering through data reduction. IEEE Trans. Fuzzy Syst. **11**(2), 262–270 (2003)
21. A.A. Esmin, R.A. Coelho, S. Matwin, A review on particle swarm optimization algorithm and its variants to clustering high-dimensional data. Artif. Intell. Rev. **44**(1), 23–45 (2015)
22. A. Fahad, N. Alshatri, Z. Tari, A. Alamri, I. Khalil, A.Y. Zomaya, . . ., A. Bouras, A survey of clustering algorithms for big data: taxonomy and empirical analysis. IEEE Trans. Emerg. Top. Comput. **2**(3), 267–279 (2014)

23. G. Hamerly, C. Elkan, Alternatives to the k-means algorithm that find better clusterings, in *Proceedings of the Eleventh International Conference on Information and Knowledge Management*, pp. 600–607 (ACM, New York, 2002)
24. J. Han, J. Pei, M. Kamber, *Data Mining: Concepts and Techniques* (Elsevier, New York, 2011)
25. Z. Huang, Clustering large data sets with mixed numeric and categorical values, in *Proceedings of the 1st Pacific-Asia Conference on Knowledge Discovery and Data Mining*, pp. 21–34 (1997)
26. Z. Huang, Extensions to the k-means algorithm for clustering large data sets with categorical values. Data Min. Knowl. Disc. **2**(3), 283–304 (1998)
27. P. Indyk, R. Motwani, Approximate nearest neighbors: towards removing the curse of dimensionality, in *Proceedings of the Thirtieth Annual ACM Symposium on Theory of Computing*, pp. 604–613 (1998)
28. T. Kanungo, D.M. Mount, N.S. Netanyahu, C.D. Piatko, R. Silverman, A.Y. Wu, An efficient k-means clustering algorithm: analysis and implementation. IEEE Trans. Pattern Anal. Mach. Intell. **24**(7), 881–892 (2002)
29. K. Krishna, M.N. Murty, Genetic K-means algorithm. IEEE Trans. Syst. Man Cybern. B Cybern. **29**(3), 433–439 (1999)
30. T. Kwok, K. Smith, S. Lozano, D. Taniar, Parallel fuzzy c-means clustering for large data sets, in *Euro-Par 2002 Parallel Processing*, pp. 27–58 (2002)
31. J.Z. Lai, T.J. Huang, Y.C. Liaw, A fast k-means clustering algorithm using cluster center displacement. Pattern Recogn. **42**(11), 2551–2556 (2009)
32. M. Laszlo, S. Mukherjee, A genetic algorithm using hyper-quadtrees for low-dimensional k-means clustering. IEEE Trans. Pattern Anal. Mach. Intell. **28**(4), 533–543 (2006)
33. Q. Li, P. Wang, W. Wang, H. Hu, Z. Li, J. Li, An efficient k-means clustering algorithm on MapReduce, in *Proceedings of Database Systems for Advanced Applications*, pp. 357–371 (2014)
34. A. Likas, N. Vlassis, J.J. Verbeek, The global k-means clustering algorithm. Pattern Recogn. **36**(2), 451–461 (2003)
35. S.A. Ludwig, MapReduce-based fuzzy c-means clustering algorithm: implementation and scalability. Int. J. Mach. Learn. Cybern. **6**, 1–12 (2015)
36. J. MacQueen, Some methods for classification and analysis of multivariate observations, in *Proceedings of the Fifth Berkeley Symposium on Mathematical Statistics and Probability* **14**(1), 281–297 (1967)
37. A. Mohebi, S. Aghabozorgi, T. Ying Wah, T. Herawan, R. Yahyapour, Iterative big data clustering algorithms: a review. Softw. Pract. Exp. **46**(1), 107–129 (2016)
38. J.D. Owens, M. Houston, D. Luebke, S. Green, J.E. Stone, J.C. Phillips, GPU computing. Proc. IEEE **96**(5), 879–899 (2008)
39. D. Pelleg, A. Moore, Accelerating exact k-means algorithms with geometric reasoning, in *Proceedings of the Fifth ACM SIGKDD International Conference on Knowledge Discovery and Data Mining*, pp. 277–281. (ACM, New York, 1999)
40. D. Pelleg, A.W. Moore, X-means: extending k-means with efficient estimation of the number of clusters, in *Proceedings of the 17th International Conference on Machine Learning*, vol. 1, pp. 727–734 (2000)
41. S.J. Phillips, Acceleration of k-means and related clustering algorithms, in *Algorithm Engineering and Experiments*, pp. 166–177 (Springer, Berlin, 2002)
42. S.J. Redmond, C. Heneghan, A method for initialising the K-means clustering algorithm using kd-trees. Pattern Recogn. Lett. **28**(8), 965–973 (2007)
43. D. Sculley, Web-scale k-means clustering, in *Proceedings of the 19th International Conference on World Wide Web*, pp. 1177–1178 (ACM, New York, 2010)
44. O. Sievert, H. Casanova, A simple MPI process swapping architecture for iterative applications. Int. J. High Perform. Comput. Appl. **18**(3), 341–352 (2004)
45. D. Singh, C.K. Reddy, A survey on platforms for big data analytics. J. Big Data **2**(1), 8 (2015)
46. M. Snir, *MPI—The Complete Reference: The MPI Core*, vol. 1 (MIT Press, Cambridge, 1998), pp. 22–56

47. A. Vattani, K-means requires exponentially many iterations even in the plane. Discret. Comput. Geom. **45**(4), 596–616 (2011)
48. T. White, *Hadoop: The Definitive Guide* (O'Reilly Media, Sebastopol, 2012)
49. R. Xu, D.C. Wunsch, Clustering algorithms in biomedical research: a review. IEEE Rev. Biomed. Eng. **3**, 120–154 (2010)
50. M. Zaharia, M. Chowdhury, M.J. Franklin, S. Shenker, I. Stoica, Spark: cluster computing with working sets. HotCloud **10**(10–10), 95 (2010)
51. A. Zayani, C.E. Ben N'Cir, N. Essoussi, Parallel clustering method for non-disjoint partitioning of large-scale data based on spark framework, in *Proceedings of IEEE International Conference on Big Data*, pp. 1064–1069 (IEEE, Piscataway, 2016)
52. J. Zhang, G. Wu, X. Hu, S. Li, S. Hao, A parallel k-means clustering algorithm with MPI, in *Proceedings of Fourth International Symposium on Parallel Architectures, Algorithms and Programming*, pp. 60–64 (2011)
53. W. Zhao, H. Ma, Q. He, Parallel k-means clustering based on MapReduce, in *Proceedings of Cloud Computing*, pp. 674–679 (2009)

Chapter 2
Overview of Efficient Clustering Methods for High-Dimensional Big Data Streams

Marwan Hassani

2.1 Introduction

Clustering is a well-established data mining concept that aims at automatically grouping *similar* data objects while separating *dissimilar* ones. This process is strongly dependent on the notion of *similarity*, which is often based on some distance measure. Thus, similar objects are usually close to each other while dissimilar ones are far from each other. The clustering task is performed without a previous knowledge of the data, or in an *unsupervised* manner.

During the early stages of data mining research, the whole data objects were considered to be statically and permanently stored in the memory. This allowed the designed data mining technique to perform as much passages over the objects as needed to deliver the desired patterns. In the era of big data, the recent growth of the data size and the easiness of collecting data made the previous settings no more convenient. The size of the continuously generated data and the limited storage capacity allow in many scenarios for a single passage over the data, and users are interested in gaining a real-time knowledge about the data as they are produced.

A data stream is an ordered sequence of objects that can be read once or very small number of times using limited processing and computing storage possibilities. This sequence of objects can be endless and flows usually at high speeds with a varying underlying distribution of the data. This fast and infinite flow of data objects does not allow the traditional permanent storage of the data and thus multiple passages are not any more possible. Many domains are dealing essentially with data streams. The most prominent examples include network traffic data, telecommuni-

M. Hassani (✉)
Analytics for Information Systems Group, Eindhoven University of Technology, Eindhoven, The Netherlands
e-mail: m.hassani@tue.nl

O. Nasraoui, C.-E. Ben N'Cir (eds.), *Clustering Methods for Big Data Analytics*, Unsupervised and Semi-Supervised Learning,
https://doi.org/10.1007/978-3-319-97864-2_2

cation records, click streams, weather monitoring, stock trading, surveillance data, health data, customer profile data, and sensor data. There is many more to come. A very wide spectrum of real-world streaming applications is expanding. Particularly in sensor data, such applications spread from home scenarios like the smart homes to environmental applications, monitoring tasks in the health sector [13], in the digital humanities using eye-tracking [21] or gesture monitoring but do not end with military applications. Actually, any source of information can easily be elaborated to produce a continuous flow of the data. Another emerging streaming data sources are social data. In a single minute, 456,000 tweets are happening, 2,460,000 pieces of content are shared on Facebook, Google conducts 3,607,080 searches, the weather channel receives 18,055,555.56 forecast requests, Uber riders take 45,787.54 trips, and 72 h of new videos are uploaded to YouTube while 4,146,600 videos are watched.[1]

Users are interested in gaining the knowledge out of these information during their same minute of generation. A delay, say till the next minute, might result in an outdated knowledge. Furthermore, with the introduction of the new regulations for data privacy protection like GDPR[2] starting into effect at the end of May 2018, businesses dealing with sensitive user profile data are not allowed anymore to store them. Thus, they would need to switch to a flexible and streaming structure to manage and analyze real-time customer behavior data.

The abovementioned emerging applications motivated dozens of research topics and development frameworks during the previous one and a half decades. Extracting a real-time knowledge out of large numbers of objects with an evolving nature required a different look at data than the traditional *static* one. Particularly, an unsupervised mining of evolving objects in the real-time was needed in multiple applications. In this chapter, we give an overview of the challenges as well as the contributions in this emerging field by highlighting some of the main algorithms there.

The remainder of this chapter is organized as follows: Sect. 2.2 introduces data streams with some applications. In Sect. 2.3, we list the challenges one has to face while designing stream clustering algorithms for mining big data. In Sect. 2.4, we present, at a high level of abstraction, recent approaches in the field of big data stream mining that overcame the challenges mentioned in Sect. 2.3. Finally, Sect. 2.5 concludes this chapter.

2.2 Streaming Data

Objects from the perspective of static mining approaches are all available in the memory while running the algorithm. As a result, the algorithm can scan objects as much as needed to generate the final output without assuming any order for reading

[1] Sources: domo.com and statisticbrain.com.

[2] https://www.eugdpr.org/.

these objects. In the streaming setting, an endless flow of objects o_1, o_2, ..., o_i, ... of the dataset D is seen at timestamps t_1, t_2, ..., t_i, ..., respectively, where $t_{i+t} > t_i$ for all i values. Each object o_i seen at t_i is a d-dimensional tuple: $o_i = (o_{i1}, o_{i2}, \ldots, o_{id}) \in D$.

Due to the endless flow of streaming objects, it is not realistic to assume a possibility of memory storage of all $o_i \in D$. This is mainly due to limitations of storage, processing power, expected response time, and even available energy. Efficient mining approaches aim at minimizing the number of scans they perform over the objects before generating the final output. Thus, both the needed storage and the required processing power are limited such that the final output that includes objects $o_i, o_{i+1}, \ldots o_j$ is generated before the arrival of object o_{j+1} at timestamp t_{j+1}.

Figure 2.1 gives some examples about real world application that produce data streams. Most of these scenarios are covered within the scope of the algorithms presented in this chapter. Figure 2.1a shows an example about wired streaming data that monitor some flowing phenomenon like network traffic data, click streams, or airport camera monitoring. Figure 2.1b presents a visualization of streaming tweets with a certain tag and within a certain time using the *Streamgraph* framework [6]. Figure 2.1c depicts an application of a wireless sensor network deployment

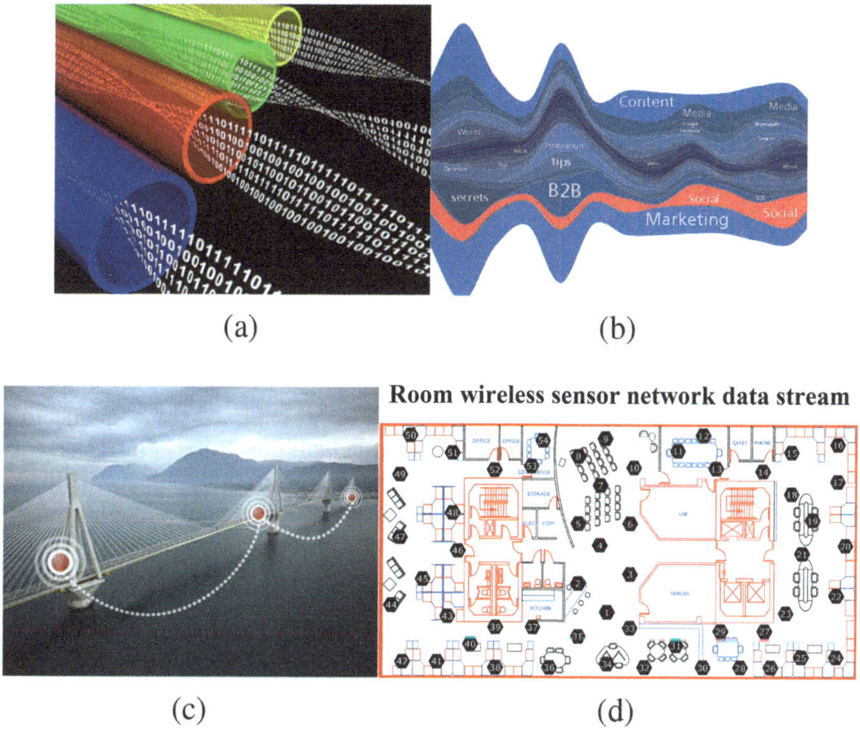

Fig. 2.1 Reference [12]. Examples of big data streams

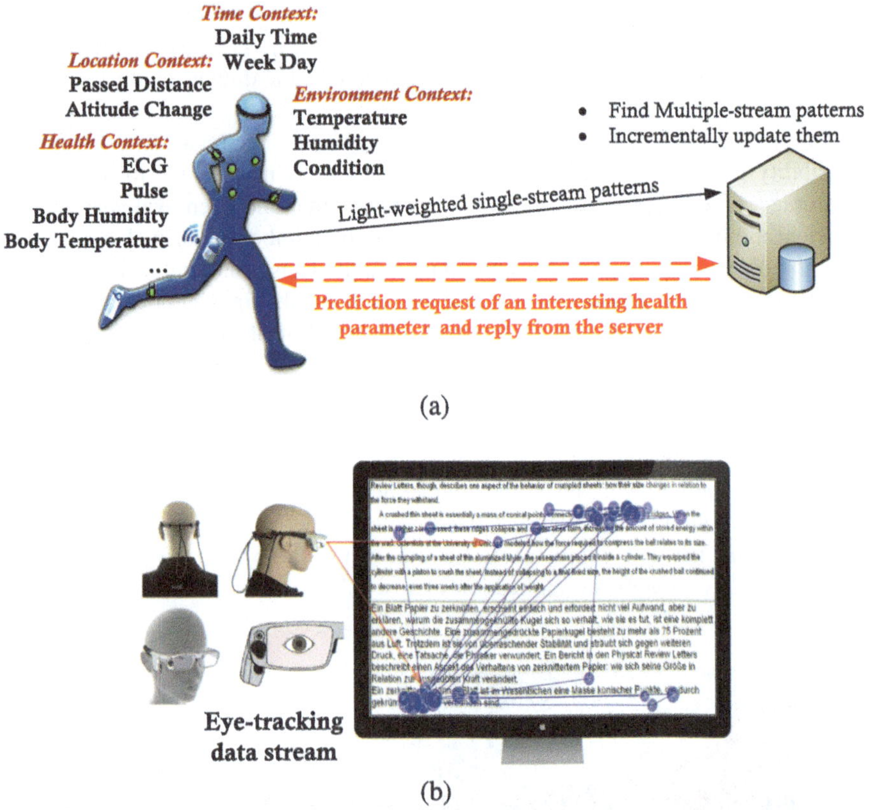

Fig. 2.2 Reference [12]. Two applications of mining body-generated streaming data. (**a**) In a health care scenario [13] and (**b**) in a translation scenario in collaboration with psycholinguists in the humanities area [21]

in a bridge for surveillance or load observation. Sensors are producing continuous streams of readings, and experts need to collect a real-time knowledge about the stability of the bridge in the case of emergency, or gather regular reports in the normal case. Similarly, Fig. 2.1d shows an example of sensors collecting temperature, humidity, and light information from multiple offices in Intel Berkeley Research Lab [10]. Such sensors are usually of limited storage, processing power, and battery life. In Fig. 2.2a, a body sensor network is producing multiple streams about the health status of the runner. Other sensors are collecting streams of other contextual information like the weather and location information. These can be processed on a local mobile device or a remote server to gain, for instance, some knowledge about the near-future status. Figure 2.2b presents another type of sensor streaming data where an eye-tracking system is used to record the duration and the position of each eye fixation over the monitor during a human reading or writing process. One task could be here finding interesting patterns that represent important correlations between eye gazes and key strokes [21].

Stream clustering aims at detecting clusters that are formed out of the evolving streaming objects. These clusters must be *continuously* updated as the stream emerges to follow the current distribution of the data. These clusters represent mainly the gained knowledge out of the clustering task. In this chapter, advanced stream clustering models are introduced. These models are mainly motivated by the basic challenges that we have observed for clustering of streaming data in real world scenarios, particularly sensor streaming data (cf. Fig. 2.1).

2.3 Challenges of Stream Clustering of Big Data

Designing stream clustering approaches has some unique special challenges. We list in the following the different paradigms that make it challenging to design a stream clustering approach.

2.3.1 *Adaptation to the Stream Changes and Outlier Awareness*

The algorithm must incrementally cluster the stream data points to detect evolving clusters over the time, while forgetting outdated data. New trends of the data must be detected at the same time of their appearance. Nevertheless, the algorithm must be able to distinguish new trends of the stream from outliers. Fulfilling the up-to-date requirement contradicts the outlier awareness one. Thus, meeting this tradeoff is one of the basic challenges of any stream clustering algorithm.

2.3.2 *Storage Awareness and High Clustering Quality*

Due to the huge sizes and high speeds of streaming data, any clustering algorithm must perform as few passages over the objects as possible. In most cases, the application and the storage limitations allow only for a single passage. However, high-quality clustering results are requested to make the desired knowledge out of the data stream. Most static clustering models tend to deliver an initial, sometimes random, clustering solution and then optimize it by revisiting the objects to maximize some similarity function. Although such multiple-passages possibility does not exist for streaming algorithms, the requirement of an optimized, high-quality clustering does still exist.

2.3.3 *Efficient Handling of High-Dimensional, Different-Density Streaming Objects*

The current huge increase of the sizes of data was accompanied with a similar boost in their number of dimensions. This applies of course to streaming data too. For such kinds of data with higher dimensions, distances between the objects grow more and more alike due to an effect termed *curse of dimensionality* [4]. According to this effect, applying traditional clustering algorithms in the full-space merely will result in considering almost all objects as outliers, as the distances between them grow exponentially with their dimensionality d. The latter fact motivated the research in the area of *subspace clustering* over static data in the last decade, which searches for clusters in all of the $2^d - 1$ subspaces of the data by excluding a subgroup of the dimensions at each step. Apparently, this implies higher complexity of the algorithm even for static data, which makes it even more challenging when considering streaming data.

Additionally, as the stream evolves, the number, the density, and the shapes of clusters may dramatically change. Thus, assuming a certain number of clusters like in k-means-based clustering models or setting a static density threshold as in the DBSCAN-based clustering models is not convenient for a stream clustering approach. A self-adjustment to the different densities of the data is strongly needed while designing a stream clustering algorithm. Again, this requirement is in conflict with the storage awareness necessity.

2.3.4 *Flexibility to Varying Time Allowances Between Streaming Objects*

An additional, natural characteristic of data streams (e.g., sensor data) is the fluctuating speed rate. Streaming data objects arrive usually with different time allowances between them, although the application settings would assume a constant stream speed. Available stream clustering approaches, called budget algorithms in this context, strongly restrict their model size to handle minimal time allowance to be on the safe side (cf. Fig. 2.6). In the case of reduced stream speed, the algorithm remains idle during the rest of the time, till the next streaming object arrives. *Anytime mining algorithms*, designed recently for static data, try to make use of any given amount time to deliver some result. Longer given times imply higher clustering quality. This idea was adopted for clustering streaming data. Although this setting can be seen as an opportunity for improving the clustering quality rather than a challenge, it is not trivial to have a flexible algorithmic model that is able to deliver some result even with very fast streams.

2.3.5 *Energy Awareness and Lightweight Clustering of Sensor Data Streams*

Wireless sensor nodes are equipped with a small processing device, a tiny memory, and a small battery in addition to the sensing unit [27]. This encouraged the research in the area of in-sensor-network mining, where the nodes do some preprocessing of the sensed data instead of simply forwarding it. In many of these applications, sensor nodes are distributed in unreachable areas without a cheap possibility of changing the battery. Thus, the usability time of the node is bounded by the battery lifetime. In this manner, besides the previously mentioned challenges, clustering sensor streaming data has to carefully consume the processing and energy resources. In fact, clustering and aggregation approaches are used within wireless sensor networks to save energy by preprocessing the data in the node, and forwarding the relevant ones merely.

2.4 Recent Contributions in the Field of Efficient Clustering of Big Data Streams

In this section, we present, at a high level of abstraction, novel, efficient stream clustering algorithms that consider all of the above challenges mentioned in Sect. 2.3. These contributions [12] are structured in the following four subsections. In Sect. 2.4.1, we present novel high-dimensional density-based stream clustering techniques. In Sect. 2.4.2, we introduce advanced anytime stream clustering approaches. In Sect. 2.4.3, we present efficient methods for clustering sensor data and aggregating sensor nodes. Finally, in Sect. 2.4.4, we present unique subspace stream clustering framework as well as the subspace cluster mapping evaluation measure. In all of the following subsections, the first and the second challenges mentioned in Sects. 2.3.1 and 2.3.2 are carefully considered. Each of the rest of the challenges (Sects. 2.3.3–2.3.5) is the main focus in one of the following subsections, as we will explain.

2.4.1 *High-Dimensional, Density-Based Stream Clustering Algorithms*

In this line of research to address big data stream clustering, we refer to three density-based stream clustering algorithms. Here, the third challenge mentioned in Sect. 2.3.3 is mainly considered.

In [18], an efficient projected stream clustering algorithm called *PreDeCon-Stream* is introduced for handling high-dimensional, noisy, evolving data streams. This technique is based on a two-phase model (cf. Fig. 2.3). The first phase repre-

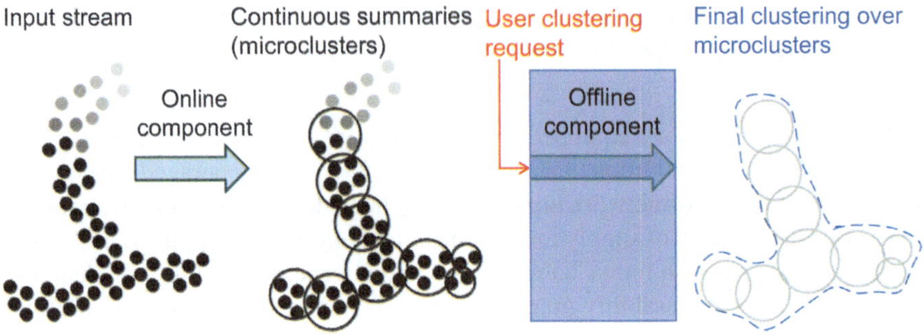

Fig. 2.3 Reference [12]. The online–offline model of stream clustering algorithms. Decayed input objects have lighter colors than recent ones

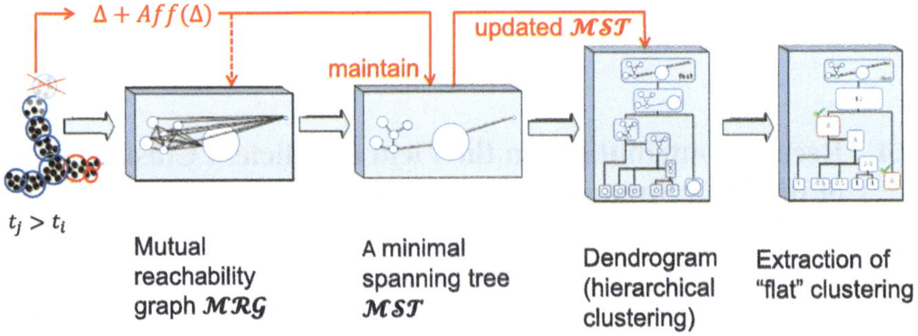

Fig. 2.4 Reference [12]. The steps of HASTREAM algorithm [19]. The incremental part is explained in red arrows to maintain the clustering at timestamp t_j after the red insertions and deletions of microclusters introduced to the old ones from timestamp t_i

sents the process of the online maintenance of data summaries, called microclusters, that are then passed to an offline phase for generating the final clustering. The technique works on incrementally updating the output of the online phase stored in a microcluster structure. Taking those microclusters that are fading out over time into consideration speeds up the process of assigning new data points to the existing clusters. The algorithm localizes the change to the previous clustering result, and smartly uses a clustering validity interval to make an efficient offline phase.

In *HASTREAM* [19], a hierarchical, self-adaptive, density-based stream clustering model is contributed (cf. Fig. 2.4). The algorithm focuses on smoothly detecting the varying number, densities, and shapes of the streaming clusters. A cluster stability measure is applied over the summaries of the streaming data (the microclusters in Fig. 2.3), to extract the most stable offline clustering. Algorithm 1 gives a pseudo-code of HASTREAM (cf. Fig. 2.4 and [19] for further details). In order to improve the efficiency of the suggested model in the offline phase,

Algorithm 1 **HASTREAM**(DataStream ds, $minClusterWeight$, bool $incUpdate$)

initialization phase
repeat
  get next point $o_i \in ds$ with current timestamp t_i;
  insert o_i in the microclusters using online parameter settings;
  if (t_i mod $updateFrequency == 0$) **then**
    if $incUpdate$ **then**
      incrementally update the minimal spanning tree MST by maintaining it;
    else
      compute the mutual reachability graph MRG and corresponding MST from scratch
    end if
    $HC \leftarrow$ extractHierarchicalClusters(MST, $minClusterWeight$);
    $C \leftarrow$ extractFlatClustering(HC);
    return C;
  end if
until data stream terminates

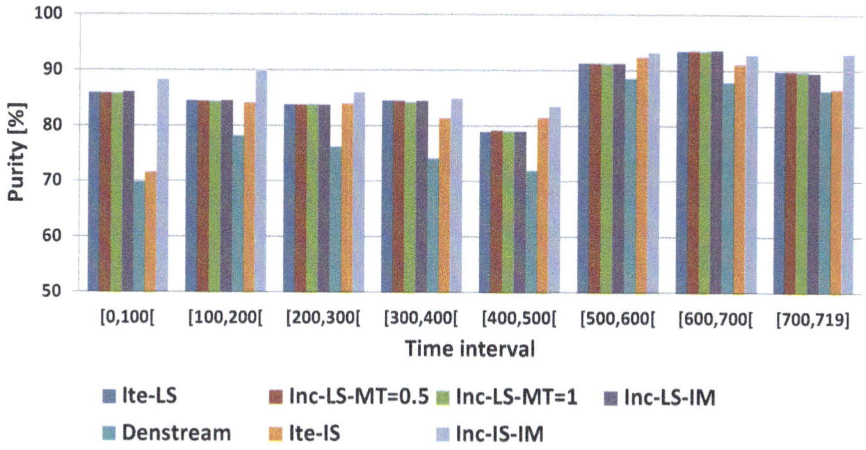

Fig. 2.5 Purity of detected clustered in five variants of HASTREAM [19] compared to DenStream [7] for the physiological dataset [26]. Timestamps are $\times 1000$

some methods from the graph theory are adopted and others were contributed, to incrementally update a minimal spanning tree of microclusters (cf. the red arrows in Fig. 2.4). This tree is used to continuously extract the final clustering, by localizing the changes that appeared in the stream, and maintaining the affected parts merely. The averaged purity for this dataset is shown in Fig. 2.5. All four variants of HASTREAM [19] have a higher averaged purity than that of DenStream [7] over the physiological dataset [26].

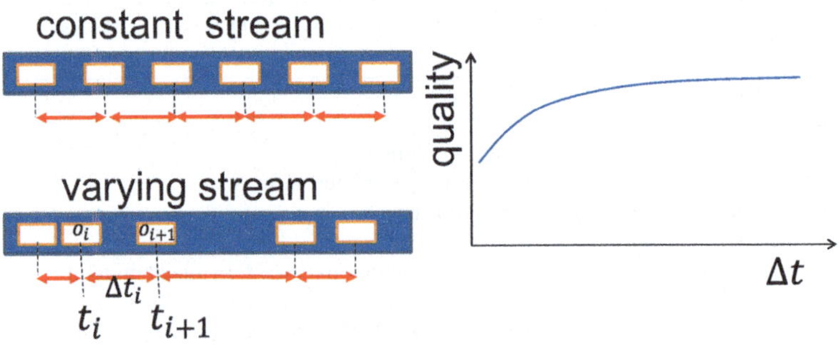

Fig. 2.6 The concept of anytime stream mining algorithms [12]

2.4.2 Advanced Anytime Stream Clustering Algorithms

By considering all other challenges, the main focus of the two algorithms presented in this section are the third and the fourth challenges mentioned in Sects. 2.3.3 and 2.3.4. Anytime algorithms build upon the realistic assumption of the varying time allowances between streaming objects (cf. Fig. 2.6). They aim at increasing the quality of their output if they were given more time (i.e., the time allowance Δt is bigger) instead of being idle as in traditional algorithms.

The *LiarTree* algorithm [17] is contributed on the online phase (cf. Fig. 2.3) to provide precise stream summaries and to effectively handle noise, drift, and novelty at any given time. It is proven that the runtime of the anytime algorithm is logarithmic in the size of the maintained model opposed to a linear time complexity often observed in previous approaches. The main contributions of this technique are enabling the anytime concept to fast adapt to the new trends of the data, filtering noise and keeping a logarithmic complexity.

In the *SubClusTree* algorithm [20], even another complexity dimension to the problem addressed in LiarTree [17] is added. The high-dimensionality paradigm of big streaming data (cf. Sect. 2.3.3) is considered together with the varying arrival times and the streaming aspects of the data (cf. Sect. 2.3.4). SubClusTree is a subspace anytime stream clustering algorithm, that can flexibly adapt to the different stream speeds and makes the best use of available time to provide a high-quality subspace clustering. It uses compact index structures to maintain stream summaries in the subspaces in an online fashion. It uses flexible grids to efficiently distinguish the relevant subspaces (i.e., subspaces with clusters) from irrelevant ones. Algorithm 2 contains a pseudo-code of SubClusTree. An object is inserted in all one-dimensional trees and if there is more time, the object is inserted into following most potential higher-dimensional tree.

In Fig. 2.7, one can obviously observe the anytime effect of SubClusTree using a 25-dimensional dataset with five clusters hidden in only 13 relevant dimensions out of the 25. For a very fast stream, a lower clustering quality is achieved. It can be seen that an average interval of 50 steps between objects is very close to the best possible value.

Algorithm 2 SubClusTree

Initialization: store subspace trees in the bitmap with bit-vector as key
repeat
    insert o_i and update the global cluster feature to time t_i;
    for $j = 1$ **to** d (number of dimensions) **do**
        insert o_{ij} into the one-dimensional tree of subspace j;
    end for
    if (t_i mod $updateFrequency == 0$) **then**
        create $candidateTrees$ and rank them according to their expected potential;
        remove trees with insufficient $potential$; *// but keep 1-dimensional trees*
    end if
    while next object o_{i+1} did not arrive yet **and** $moreTreesAvailable$ **do**
        insert o_i into next subspace tree
    end while
until data stream terminates

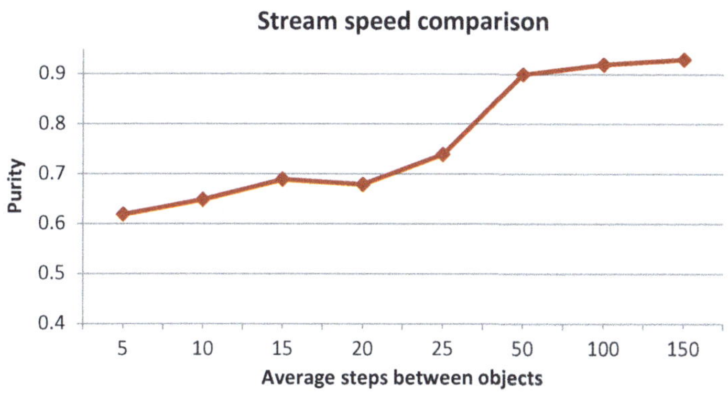

Fig. 2.7 Reference [12]. Average purity achieved by varying the inter-arrival times of the objects Δt

2.4.3 Energy-Efficient Algorithms for Aggregating and Clustering Sensor Streaming Data

In the three algorithms contributed in this line of research, the main focus is the fifth challenge mentioned in Sect. 2.3.5, while keeping the main challenges from Sects. 2.3.1 and 2.3.2 in mind.

The *EDISKCO* algorithm [15] is an Energy-Efficient Distributed In-Sensor-Network K-Center Clustering algorithm with Outliers. Sensor networks have limited resources in terms of available memory and residual energy. As a dominating energy consuming task, the communication between the node and the sink has to be reduced for a better energy efficiency. Considering memory, one has to reduce the amount of stored information on each sensor node. EDISKCO performs an outlier-aware k-center clustering [11] over the sensed streaming data in each node and forwards the clustering result to the neighboring coordinator node. For that,

Algorithm 3 EDISKCO on Node side

1: **Initialization:** Select the first *InitPts* objects from the stream;
2: perform an offline k-center clustering over the *InitPts* objects;
3: send the k centers and the cluster radius R to the coordinator
4: **repeat**
5: insert o_i in available clusters and update cluster centers $c_1, c_2, \ldots c_k$;
6: if changed, send the new centers to the coordinator;
7: **if** o_i does not fit in any available cluster **and** number of clusters $== k$ **then**
8: recluster after getting the coordinator acknowledgment;
9: **end if**
10: **until** data stream terminates

Algorithm 4 EDISKCO on Coordinator side

1: **repeat**
2: receive k-center solutions from all nodes and perform another k-center clustering on them;
3: Acknowledge reclustering by broadcasting the biggest current radius from all nodes;
4: Change coordinator w.r.t. residual energy;
5: **until** data stream terminates

each node consumes considerably less energy than the cost of forwarding all of its readings to the sink. The update from the node to the coordinator happens only upon a certain deviation of the cluster radii, controlled by an ϵ threshold, or upon the birth or the deletion of a cluster. One node is selected as a coordinator from each spatially correlated subgroup of the nodes, depending on the amount of the residual energy. The coordinator performs another k-center clustering over the multiple clustering solutions arriving from its neighboring nodes and forwards the solution to the far sink. These are major contributions as one can perform a single passage over the data by using a $O(k)$ storage over the nodes and getting finally a high clustering quality of a $(4 + \epsilon)$-approximation to the optimal clustering. But, the main contribution is performing up to 95% less communication tasks on the nodes compared to a state-of-the-art technique. Thus, huge savings of energy are achieved. Algorithm 3 summarizes the node side procedure of EDISKCO while Algorithm 4 abstracts the approach on the coordinator side.

A weighted version of EDISKCO, called *SenClu*, is contributed in [14]. It guarantees a faster adaptation to the new trends of the drifting data streams. The technique gives more importance to new data points, while slowly forgetting older ones by giving them less weight. To achieve this, a novel, light-weighted decaying function is contributed, that can be implemented on the tiny processing unit and works on the limited storage capacity of sensor nodes. SenClu achieves even better clustering quality than EDISKCO while draining almost the same amount of energy.

One can see from Table 2.1 that on the real dataset (i9-Sensor Dataset), SenClu [14] consumes less than two Joules more than EDISKCO [15], and absorbs considerably less energy than the state-of-the-art competitor PG [8]. When using the Physiological Sensor Dataset [26], SenClu consumes less energy than both competitors. Figure 2.8a shows that SenClu [14] and EDISKCO [15] always have a better clustering quality than PG [8] on the node side. Because PG is more sensitive

Table 2.1 Average energy consumption in Joule of a single node in the network by the end of each dataset when using SenClu, EDISKCO, and PG

Dataset	Size	Nodes	SenClu [14]	EDISKCO [15]	PG [8]
i9-Sensor	40,589	19	28,770.2	**28,768.38**	28,792.5
Physio [26]	24,000	12	**17,074.3**	17,074.4	17,078.9

Lowest energy consumption is in bold

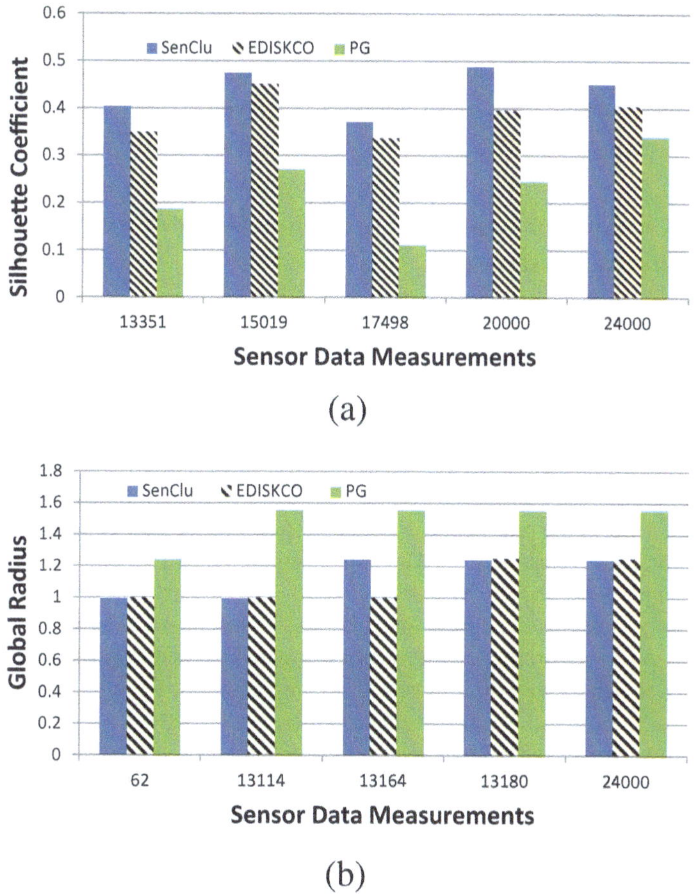

Fig. 2.8 Reference [12]. The clustering quality using the Real Physiological Sensor Dataset [26] over different parts of the input stream data. (**a**) Silhouette coefficient (higher is better), (**b**) R_{global} (lower is better)

to noise than SenClu and EDISKCO, it is performing considerably worse than the others on this relatively noisy dataset. Figure 2.8b is showing that on the node side, SenClu is having most of the time the same global radius as EDISKCO. Only for a short time, SenClu is having a bigger radius than EDISKCO.

A further challenge for aggregating streaming data within the sensor network is tackled. The physical clustering of sensor nodes depending on their similarity is considered in the presented *ECLUN* algorithm [16]. The readings of a carefully selected representative node are used to simulate the measurements of similar nodes. While the recent approaches concentrated on the full-dimensionality correlation between the readings, ECLUN selects the representatives depending on the subspace correlation between some attributes of the measurements as well as the spatial similarity. Additionally, the usage of energy between the nodes is uniformly distributed, and, thus, the cases of single-node clusters are handled by changing representatives according to the residual energy. This results in a longer lifetime of the whole sensor network as nodes die close to each other.

2.4.4 A Framework and an Evaluation Measure for Subspace Stream Clustering

This section presents some contributions mainly in the *Evaluation and Visualization* step of the KDD (Knowledge Discovery in Databases) process.

The first subspace clustering evaluation framework over data streams, called *Subspace MOA*, is presented in [22]. This open-source framework is based on the MOA stream mining framework [5], and has three phases (cf. Fig. 2.9). In the online phase, users have the possibility to select one of three most famous summarization techniques to form the microclusters. Upon a user request for a final clustering, the regeneration phase constructs the data objects out of the current microclusters. Then, in the offline phase, one of five subspace clustering algorithms can be selected. In addition to the previous combinations, the framework contains available projected stream clustering algorithms like PreDeConStream [18] and HDDStream [25]. The framework is supported with a subspace stream generator, a visualization interface, and various subspace clustering evaluation measures. With the increase of the size of high-dimensional data, applying traditional subspace clustering algorithms is impossible due to their exponential complexities. Figure 2.10 shows, for instance,

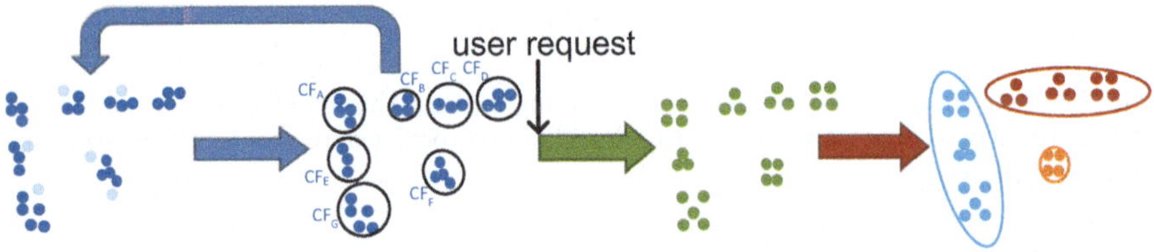

Fig. 2.9 Reference [12]. *Subspace MOA* model for stream subspace clustering for big data. The blue arrows represent the online phase, the green arrow represents the regeneration phase, and the red arrow represents the offline phase

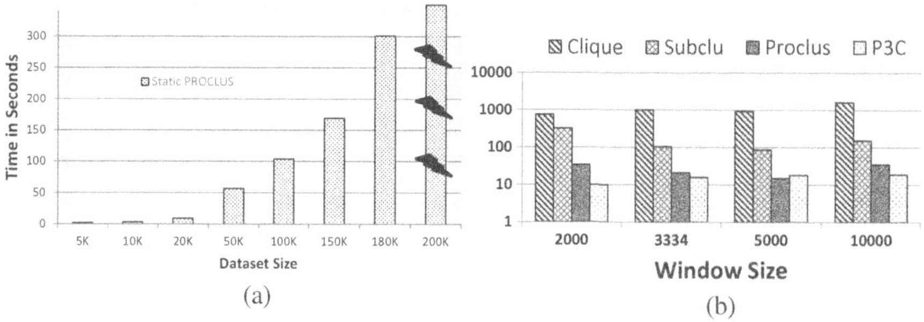

Fig. 2.10 Reference [12]. (**a**): The runtime of the *static* subspace clustering algorithm: PROCLUS [1]. Beginning from a sub-dataset size of 200 K objects only, the algorithm fails to successfully finish the running. (**b**): A successful run of PROCLUS (and other subspace clustering algorithms Clique [3], SubClu [23], and P3C [24]) over the whole KDD dataset [9] when applying the *streaming* PROCLUS using Subspace MOA with CluStream [2] in the online phase on the same machine

that when using the same machine, it was only possible by using Subspace MOA, to get a relatively large dataset clustered with a subspace clustering algorithm (PROCLUS [1]).

In [22], a novel external evaluation measure for stream subspace clustering algorithms called *SubCMM*: Subspace Cluster Mapping Measure is contributed. SubCMM is able to handle errors caused by emerging, moving, or splitting subspace clusters. This first evaluation measure that is designed to reflect the quality of stream subspace algorithms is directly integrated in the Subspace MOA framework. This new measure was extensively compared against state-of-the-art full-space stream clustering evaluation measures. The experimental evaluation, performed using the Subspace MOA framework, depicts the ability of *SubCMM* to reflect the different changes happening in the subspaces of the evolving stream.

2.5 Conclusion

In this chapter, we have mainly addressed the following three v's of big data: velocity, volume, and variety. Various streaming data applications, with huge volumes and varying velocities, were considered in different scenarios and applications.

A list of the recent challenges that face the designer of a stream clustering algorithm was shown and deeply discussed. Finally, some recent contributions on four main research lines in the area of big data stream mining were presented and their fulfillment to the design requirements was highlighted. Table 2.2 summarizes the main properties of some stream clustering algorithms mentioned in this chapter.

Table 2.2 Properties of the different stream clustering algorithms discussed in this chapter (*na* = not applicable, ∼ = indirectly available)

Algorithm	Online–offline model	Density-based (in online phase)	Density-based (in offline phase)	Hierarchical	Outlier-aware (cf. Sect. 2.3.1)	Density-adaptive (cf. Sect. 2.3.3)	Anytime (cf. Sect. 2.3.4)	Lightweight clustering (cf. Sect. 2.3.5)	Incremental (in offline phase)	Overlapping clusters and subspaces	Data structure of micro-clusters
Full-space stream clustering algorithms											
CluStream [2]	✓									*na*	*Pyramidal time frame*
DenStream [7]	✓	✓	✓		✓					*na*	*List*
LiarTree [17]	✓	✓	*na*		✓		✓			*na*	*Index*
HASTREAM [19]	✓	✓	✓	✓	✓	✓				*na*	*Index* or *List*
EDISKCO [15]	∼			∼	✓			✓		*na*	*List*
SenClu [14]	∼			∼	✓			✓	✓	*na*	*List*
Subspace stream clustering algorithms											
PreDeConStream [18]	✓	✓	✓		✓						*List*
HDDStream [25]	✓	✓	✓		✓						*List*
SubClusTree [20]	✓	✓	*na*		✓		✓			✓	*Index*
ECLUN [16]	∼			∼	✓			✓	✓	*na*	*List*

References

1. C.C. Aggarwal, J.L. Wolf, P.S. Yu, C. Procopiuc, J.S. Park, Fast algorithms for projected clustering, in *Proceedings of the 1999 ACM SIGMOD International Conference on Management of Data*, SIGMOD '99, pp. 61–72 (ACM, New York, 1999)
2. C.C. Aggarwal, J. Han, J. Wang, P.S. Yu, A framework for clustering evolving data streams, in *Proceedings of the 29th International Conference on Very Large Data Bases*, VLDB '03, pp. 81–92 (VLDB Endowment, Los Angeles, 2003)
3. R. Agrawal, J. Gehrke, D. Gunopulos, P. Raghavan, Automatic subspace clustering of high dimensional data for data mining applications, in *Proceedings of the 1998 ACM SIGMOD International Conference on Management of Data*, SIGMOD '98, pp. 94–105 (ACM, New York , 1998)
4. K.S. Beyer, J. Goldstein, R. Ramakrishnan, U. Shaft, When is "nearest neighbor" meaningful? in *Proceedings of the 7th International Conference on Database Theory*, ICDT '99, pp. 217–235 (Springer, Berlin, 1999)
5. A. Bifet, G. Holmes, R. Kirkby, B. Pfahringer, MOA: Massive online analysis. J. Mach. Learn. Res. **11**, 1601–1604 (2010)
6. L. Byron, M. Wattenberg, Stacked graphs - geometry & aesthetics. IEEE Trans. Vis. Comput. Graph. **14**(6), 1245–1252 (2008)
7. F. Cao, M. Ester, W. Qian, A. Zhou, Density-based clustering over an evolving data stream with noise, in *Proceedings of the 6th SIAM International Conference on Data Mining*, SDM '06, pp. 328–339 (2006)
8. G. Cormode, S. Muthukrishnan, W. Zhuang, Conquering the divide: continuous clustering of distributed data streams, in *IEEE 23rd International Conference on Data Engineering*, ICDE '07, pp. 1036–1045 (IEEE Computer Society, Washington, 2007)
9. N.I. Dataset, KDD cup data (1999). http://kdd.ics.uci.edu/databases/kddcup99/kddcup99.html
10. I. Dataset, Dataset of Intel Berkeley Research Lab (2004) URL: http://db.csail.mit.edu/labdata/labdata.html
11. S. Guha, Tight results for clustering and summarizing data streams, in *Proceedings of the 12th International Conference on Database Theory*, ICDT '09, pp. 268–275 (ACM, New York, 2009)
12. M. Hassani, Efficient clustering of big data streams, Ph.D. thesis, RWTH Aachen University, 2015
13. M. Hassani, T. Seidl, Towards a mobile health context prediction: sequential pattern mining in multiple streams, in *Proceedings of the IEEE 12th International Conference on Mobile Data Management*, vol. 2 of *MDM '11*, pp. 55–57 (IEEE Computer Society, Washington, 2011)
14. M. Hassani, T. Seidl, Distributed weighted clustering of evolving sensor data streams with noise. J. Digit. Inf. Manag. **10**(6), 410–420 (2012)
15. M. Hassani, E. Müller, T. Seidl, EDISKCO: energy efficient distributed in-sensor-network k-center clustering with outliers, in *Proceedings of the 3rd International Workshop on Knowledge Discovery from Sensor Data*, SensorKDD '09 @KDD '09, pp. 39–48 (ACM, New York, 2009)
16. M. Hassani, E. Müller, P. Spaus, A. Faqolli, T. Palpanas, T. Seidl, Self-organizing energy aware clustering of nodes in sensor networks using relevant attributes, in *Proceedings of the 4th International Workshop on Knowledge Discovery from Sensor Data*, SensorKDD '10 @KDD '10, pp. 39–48 (ACM, New York, 2010)
17. M. Hassani, P. Kranen, T. Seidl, Precise anytime clustering of noisy sensor data with logarithmic complexity, in *Proceedings of the 5th International Workshop on Knowledge Discovery from Sensor Data*, SensorKDD '11 @KDD '11, pp. 52–60 (ACM, New York, 2011)
18. M. Hassani, P. Spaus, M.M. Gaber, T. Seidl, Density-based projected clustering of data streams, in *Proceedings of the 6th International Conference on Scalable Uncertainty Management*, SUM '12, pp. 311–324 (2012)
19. M. Hassani, P. Spaus, T. Seidl, Adaptive multiple-resolution stream clustering, in *Proceedings of the 10th International Conference on Machine Learning and Data Mining*, MLDM '14, pp. 134–148 (2014)

20. M. Hassani, P. Kranen, R. Saini, T. Seidl, Subspace anytime stream clustering, in *Proceedings of the 26th Conference on Scientific and Statistical Database Management*, SSDBM' 14, p. 37 (2014)
21. M. Hassani, C. Beecks, D. Töws, T. Serbina, M. Haberstroh, P. Niemietz, S. Jeschke, S. Neumann, T. Seidl, Sequential pattern mining of multimodal streams in the humanities, in *Datenbanksysteme für Business, Technologie und Web (BTW), 16. Fachtagung des GI-Fachbereichs "Datenbanken und Informationssysteme" (DBIS), 4.-6.3.2015 in Hamburg, Germany. Proceedings*, pp. 683–686 (2015)
22. M. Hassani, Y. Kim, S. Choi, T. Seidl, Subspace clustering of data streams: new algorithms and effective evaluation measures. J. Intell. Inf. Syst. **45**(3), 319–335 (2015)
23. K. Kailing, H.-P. Kriegel, P. Kröger, Density-connected subspace clustering for high-dimensional data, in *Proceedings of the SIAM International Conference on Data Mining*, SDM '04, pp. 246–257 (2004)
24. G. Moise, J. Sander, M. Ester, P3C: a robust projected clustering algorithm, in *Proceedings of the 6th IEEE International Conference on Data Mining*, ICDM '07, pp. 414–425 (IEEE, Piscataway, 2006)
25. I. Ntoutsi, A. Zimek, T. Palpanas, P. Kröger, H.-P. Kriegel, Density-based projected clustering over high dimensional data streams, in *Proceedings of the 12th SIAM International Conference on Data Mining*, SDM '12, pp. 987–998 (2012)
26. Physiological dataset. http://www.cs.purdue.edu/commugrate/data/2004icml/
27. J. Polastre, R. Szewczyk, D. Culler, Telos: enabling ultra-low power wireless research, in *Proceedings of the 4th International Symposium on Information Processing in Sensor Networks*, IPSN '05 (IEEE Press, , Piscataway, 2005)

Chapter 3
Clustering Blockchain Data

Sudarshan S. Chawathe

3.1 Introduction

Blockchains, cryptocurrencies, and Bitcoin in particular, have grown increasingly popular by most social and financial metrics [12]. They receive wide coverage in mainstream news outlets and publications. Perhaps the metric that garners most attention is the appreciation of several cryptocurrencies relative to conventional fiat currencies such as the US dollar. Figure 3.1 summarizes this appreciation (note logarithmic scale). While the valuations and their longevity are subject to debate and speculation [23], there is much less uncertainty about the underlying blockchain technology and distributed socio-technical model, which seem here to stay [52, 61]. For instance, several major financial institutions are deploying private or restricted blockchains for internal or co-institutional use, and there is a growing collection of financial and non-financial applications [42]. Another example of non-cryptocurrency use of blockchains is a system for publicizing scientific datasets [48]. In their overview of Bitcoin in the context of computer science research, Bonneau et al. conclude [10]:

> Our extensive analysis of Bitcoin based on both the academic and (vast, fragmented) online literature shows a renaissance of new ideas in designing a practical cryptocurrency, a longstanding challenge for the computer security community. Innovation has not been limited to new cryptocurrency protocol designs but has touched many areas of computer security, distributed systems, hardware design, and economics. This is a rich and deep space and it should not be overlooked simply because many ideas did not originate from traditional computer science research institutes.

S. S. Chawathe (✉)
University of Maine, Orono, ME, USA
e-mail: chaw@eip10.org

O. Nasraoui, C.-E. Ben N'Cir (eds.), *Clustering Methods for Big Data Analytics*, Unsupervised and Semi-Supervised Learning,
https://doi.org/10.1007/978-3-319-97864-2_3

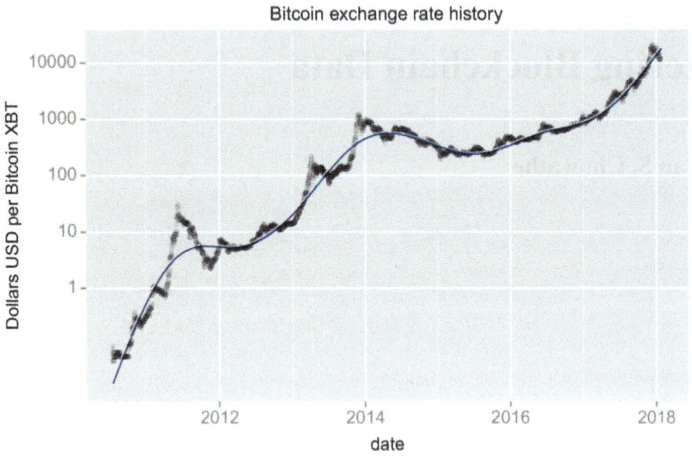

Fig. 3.1 Bitcoin (XBT) prices in US dollars (USD), *log scale*, by date. Plot generated by combining historical exchange rate data from a freely available spreadsheet [5] (2010–2013, from the defunct Mt. Gox exchange) with similar data from *CoinMarketCap* [13] (2013–2018, from a weighted average of several markets)

All the above makes a strong case for studying blockchains in general from various perspectives, including social, economic, networks and distributed systems, data mining, and more. The topic of this chapter is much more narrow: the analysis of blockchain data, in particular clustering and related methods.

3.1.1 Motivation

We may classify the motivations for clustering and other related analysis of blockchains into the following three categories.

3.1.1.1 Fraud Detection and Law Enforcement

Bitcoin has witnessed some high profile fraud and thefts, such as the theft of currency from exchanges or wallet sites. By design, all transfer of currency is publicly and persistently recorded, so it would seem that thieves should be exposed when they try to cash out by exchanging the cryptocurrency for conventional fiat currency [34]. A related objective is tracing the flow of finances among people and organizations engaged in criminal activities and using Bitcoin for payments (e.g., the *WannaCry* ransomware attack [14]). However such detection is difficult in practice because there is no requirement for Bitcoin addresses to be registered in any manner, at least until a transaction with a conventional currency exchange is

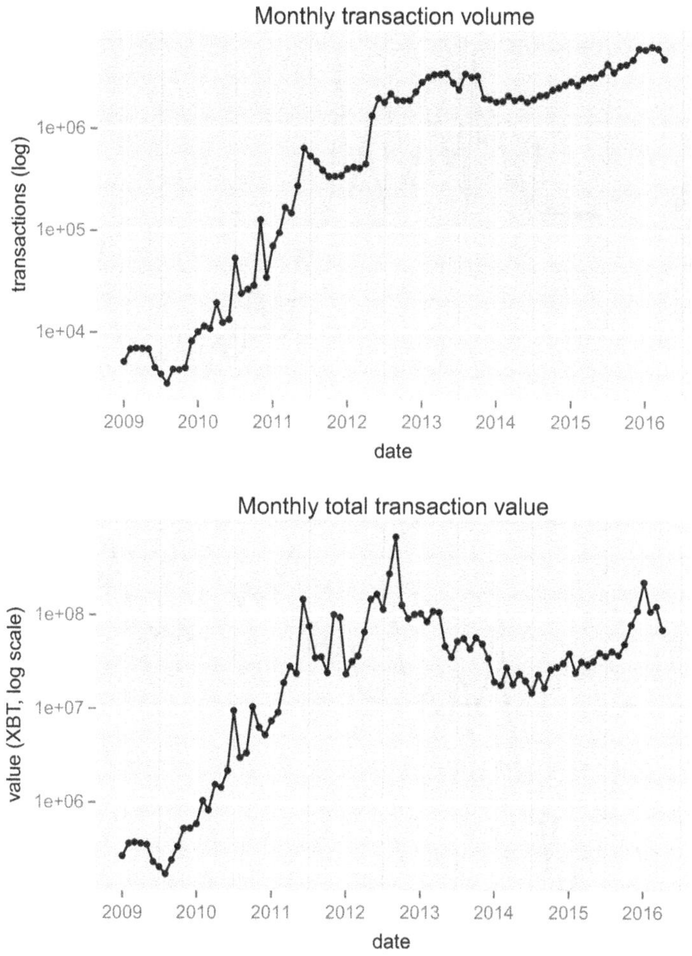

Fig. 3.2 Historical plots of the number of transactions per month (upper) and total transacted currency (Bitcoin) per month (lower). Note the logarithmic scale on both vertical axes. Further, the lower plot should be interpreted in the context of rapidly increasing XBT/USD exchange rate (Fig. 3.1). Plots generated using blockchain data from the Bitcoin Core client

required. Further, there is no practical limit on the number of addresses that a person or organization may generate, so it is easy to create complex chains of transfers that obfuscate the origin of funds. Nevertheless, a careful study of Bitcoin addresses and transactions, along with some external information, such as accidentally or casually revealed connections between Bitcoin addresses and entities, can be used to uncover many such schemes [27] (Figs. 3.2 and 3.3).

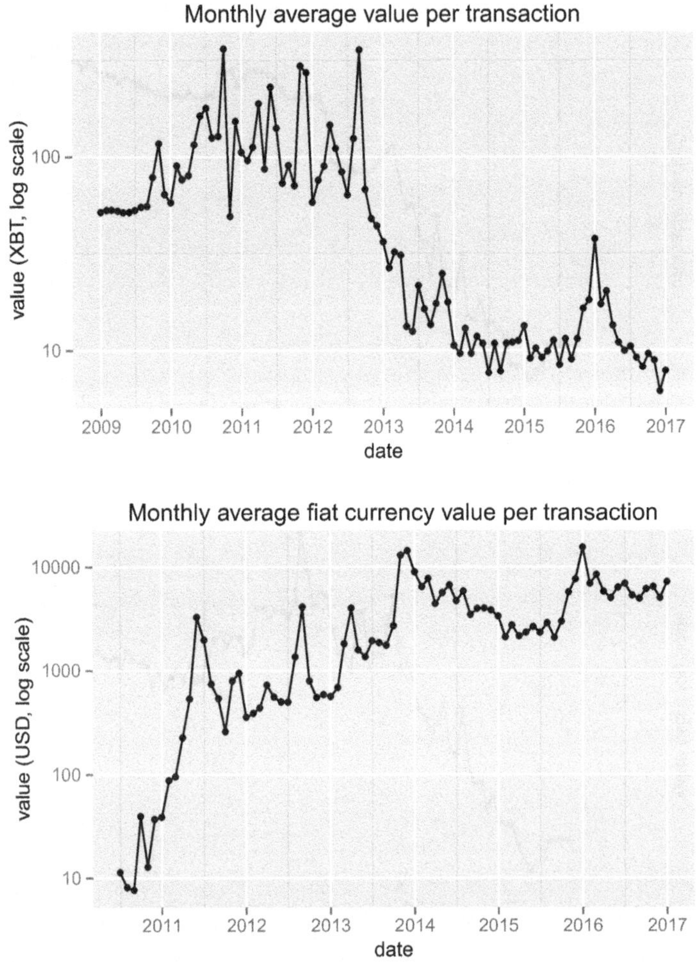

Fig. 3.3 Historical plots of the value traded per transaction, averaged monthly, in Bitcoin (XBT, upper) and US dollar (USD, lower) units. Note the logarithmic scale on both vertical axes. Plots generated with blockchain data from the Bitcoin Core client

3.1.1.2 Systems Insights

The global, distributed, peer-to-peer system that maintains a blockchain is an intriguing and impressive artifact in its own right, worthy of study for computer science and, in particular, systems insights [2, 50]. More specifically, by studying the frequency, value, and other characteristics of transactions, and other properties of the underlying system, and designers of not only blockchain systems, but other large distributed systems as well, are likely to derive valuable insights. The permanently recorded, globally consistent history of transactions and timestamps spanning several years provides a rare opportunity for longitudinal studies in this context [16]. For example, by examining transaction and block timestamps, and the volume of concurrent transactions, one may be able to make inferences about the response of the system to certain load conditions. In a complementary manner, the voluminous data generated by blockchains, and the particular needs of their analysis, provide good environments for testing systems, especially their scalability [28].

3.1.1.3 Anonymity and Traceability

Bitcoin provides a form of anonymity that has been called *pseudoanonymity*: On one hand, every Bitcoin transaction is fully public, and permanent, and thus practically impossible to conceal. On the other, the only form of identity required to transact in Bitcoin is a randomly generated address, which on its own lacks any identifying information. While this anonymity can and has been used for nefarious purposes, it is also useful for legitimate privacy reasons. However, as noted in the original Bitcoin paper [43] itself, the anonymity of a Bitcoin address's owner may be compromised by patterns in transactions. Studying clustering and related methods for analyzing the blockchain allows the community to better understand the limitations of the anonymity provided by Bitcoin and other blockchains [3, 40, 45].

3.1.2 Contribution

The main contributions of this chapter may be summarized as follows.

- **Blockchain fundamentals:** While blockchain technology and its major deployments in cryptocurrency and other applications have been in place for several years, it is still very difficult to find concise and accurate descriptions of the key data, protocols, and related mechanisms. Research publications on problems and applications motivated by blockchains (e.g., clustering) typically use an abstract model (e.g., transaction graphs) without describing the surrounding technology in much detail. Conversely, publications targeting blockchain developers describe the implementations in great detail but typically do not devote much attention to abstracting out the key concepts and problems that may be of interest to researchers. Combined with the rapidly changing nature of many details, these factors make it very difficult for researchers interested in blockchains (but not familiar with them) to enter the field. An important contribution of this chapter is a description of blockchain data at a level of detail appropriate for such research purposes. In particular, it bridges the gap between low-level developer-style documentation and fully abstract descriptions of subproblems arising from blockchain data (Fig. 3.4).
- **Models for blockchain data:** In addition to the overview of blockchain data covered as part of the above description, this chapter describes alternatives for

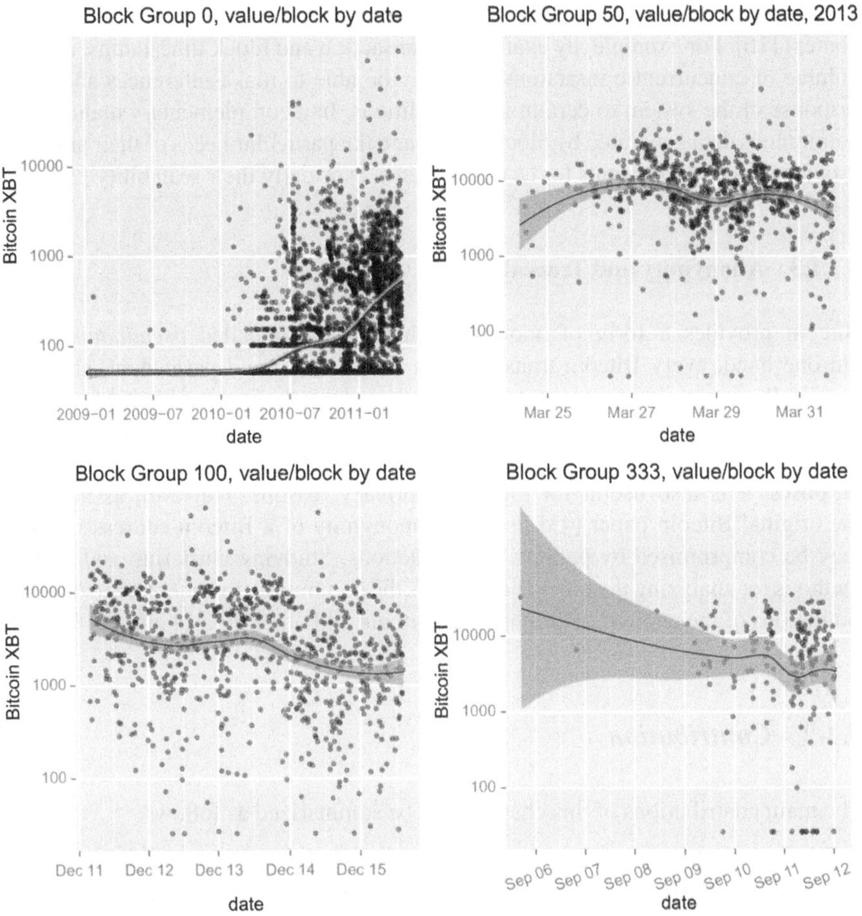

Fig. 3.4 Bitcoin (XBT) value per block (of the blockchain), by date, for different block groups (as implemented by the Bitcoin Core client). Data for block group 0 has been subsampled

modeling such data at a conceptual level. Some of these models, such as those based on the transaction and node graphs, are common in prior work but they are presented here along with the details of how they connect with the lower-level concepts in the block chain. In addition, ways of using other, less commonly used, data from the blockchain are also discussed.

- **Methods to cluster blockchain data:** Clustering methods are described in the context of the above conceptual models. Two categories of methods are described. The first consists of applying well-known clustering algorithms to features extracted from the blockchain models. The focus for such methods is on feature extraction methods. The second category of methods are more specialized to blockchains and, in particular, to the important task of address merging based on patterns in the blockchain.

- **Evaluation metrics:** Evaluating the quality of the output of clustering algorithms on blockchain data is a difficult task due to the lack of test datasets. This chapter outlines several promising criteria that may be used for this purpose, including criteria based on intrinsic metrics and the Mahalanobis distance.

3.1.3 Organization

Section 3.1 outlined the context and motivations for studying blockchain data. Blockchains, and in particular the key aspects of the data they generate, are described in Sect. 3.2. These data may be modeled in diverse and complementary ways for the purposes of clustering and related analysis; these models are described in Sect. 3.3. Clustering methods using these models are presented in Sect. 3.4, and techniques for their evaluation are described in Sect. 3.5, before concluding in Sect. 3.6.

3.2 Blockchain Data

Documentation of data from blockchains tends to be fluid and difficult to grasp, in contrast to the stable, documented datasets found in many other application domains of clustering. One reason for this difference is the recency of widespread use of blockchains and the resulting growth of associated data. Another is the consensus-based development and operations model that drives the processes generating this data. In particular, one unconventional aspect of Bitcoin is that there is no official specification, formal or informal. While there is certainly some detailed documentation of the protocol and implementation that documentation is explicit in stating:

> The Bitcoin.org Developer Documentation describes how Bitcoin works to help educate new Bitcoin developers, but it is not a specification—and it never will be. [...]
> The only correct specification of consensus behavior is the actual behavior of programs on the network which maintain consensus. As that behavior is subject to arbitrary inputs in a large variety of unique environments, it cannot ever be fully documented here or anywhere else.

It is therefore especially important to develop a shared understanding and effective abstraction of blockchain data. For concreteness, the description that follows focuses on the Bitcoin blockchain and the Bitcoin Core (Satoshi client) reference implementation [10]. However, many of the key ideas, such as transactions, addresses, and the flow of currency, are also applicable to other blockchains, such as Ethereum [11]. Our focus in this section is on providing a description of the Bitcoin blockchain, and in particular its key data structures, that is faithful to the operational blockchain. While simplification and elision of details are necessary in order to keep the presentation manageable, those simplifications and omissions do not substantially affect the data-centric view that is needed for clustering and related analyses.

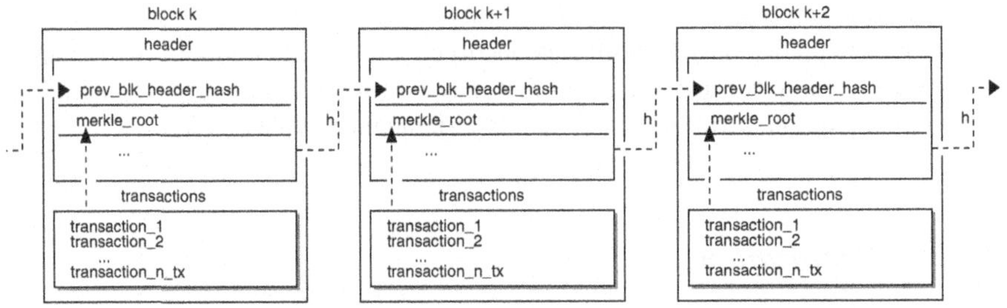

Fig. 3.5 A simplified view of the blockchain. The transactions in each block are combined pairwise recursively in tree form to yield the Merkle-root hash in the block header. Each block's header is hashed and included in the following block

3.2.1 Blocks

The core idea of the blockchain is extremely simple and is illustrated in Fig. 3.5. Intuitively, information is divided into blocks (approximately 1 MB each) and the blocks are chained together by cryptographically secure hashes: Each block includes the hash of its predecessor, and its own hash in turn is included by its successor. The primary effect of such chaining is that it is practically impossible to modify information ("rewrite history") once a block has been part of the chain for some time and has a few successors. The reason is that making a change in a block would change its hash (barring a major cryptographic breakthrough), which in turn modifies its successor blocks, recursively, up to the most recent block. Thus, making a single, small, change entails changing a very large number of blocks, which is not possible for an isolated actor or small group in the large peer-to-peer network. (Attacks in which a colluding group manages to gain control of a majority of the network peers are still possible.)

More concretely, the primary contents of each block, i.e., the transactions, are not hashed en masse with the rest of the block. Rather, they are arranged in a *Merkle tree* [4, 39] structure to yield an aggregate hash of the transactions, and this hash is included in the header of the block along with some other bookkeeping information. Further, only the header of each block (not the rest, which holds the transactions) is directly hashed for the purposes of chaining hashes as described above. Figure 3.6 illustrates the manner in which transactions are hierarchically hashed to yield the hash that is included in the block header. The major benefit of computing the hash of the set of transactions in this manner, instead of a more direct hash of the concatenated transactions, is that verifying that a transaction is part of a block (and thus the blockchain) requires only $\Theta(\log t)$ computational work, where t is the number of transactions in the block, in contrast to $\Theta(t)$ for the more direct method.

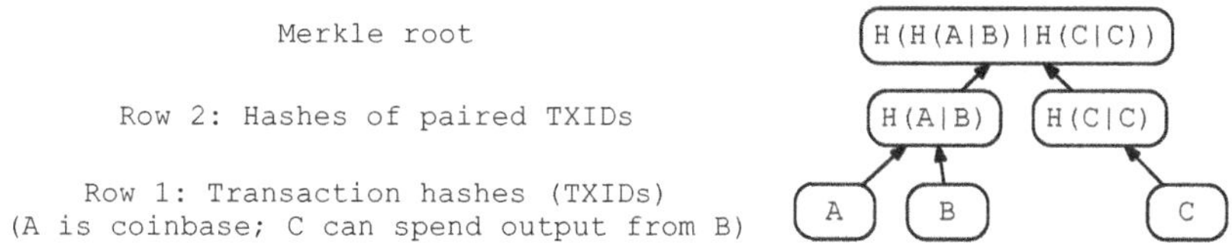

Fig. 3.6 Computation of Merkle-tree hashes for a sequence of transactions. (Figure adapted from the bitcoin.org source files, under the MIT License.) The hash function is $H(x) = \text{sha256}(\text{sha256}(x))$

3.2.2 Mining

The task of assembling transactions and other information into a block that meets the above, and other, requirements imposed (using consensus) by the Bitcoin blockchain is performed by persons or groups called *miners*. (Terms such as builder or architect are more accurate, but miner is firmly entrenched.) In order to make it computationally very difficult to modify confirmed blocks in the blockchain, an important requirement in assembling a block is that the hash of the block have a numerical value lower than a network-specified (consensus) threshold. Achieving this goal entails determining a suitable value of a specially designated field, called the *nonce*.

In order to incentivize miners to perform this computationally (and financially) expensive operation, the blockchain awards (again, by consensus) the miner of each block with a predetermined amount of Bitcoin, called the *block subsidy*. Further, in order to incentivize miners to include transactions (which is not, strictly, a requirement for receiving the block subsidy), transaction originators also include a *transaction fee*. The combination of these two incentives (block subsidy and transaction fee) is called the *block reward*, and it is what keeps the peer-to-peer blockchain network going. By design, the Bitcoin network increases the ratio of transaction fees to block subsidy over time so that, in the early phases, the block reward was composed entirely of the subsidy while in the future, it will be composed entirely of transaction fees.

3.2.3 Transactions

Each transaction transfers all the currency from one or more *inputs* to an ordered list of one or more *outputs*. (The special *coinbase* transaction, used to award currency to the block miner, is the only kind, and only one per block, that has no inputs.) The ordering of the outputs is important because a specific output in a transaction is identified (by an input in a subsequent transaction) by providing the index of the desired output in this ordering, along with the *transaction identifier*, which is a hash (SHA256, twice) of the raw transaction data.

Each output of a transaction may be thought of as a record with two fields: The first is *value*, which is the number of *satoshi* (10^{-8} Bitcoin) that are to be transferred to that output. The second is a *pubkey script* that determines the requirements imposed on anyone who wishes to spend these satoshis, by conceptually using this output as an input for a subsequent transaction. (We skip the cryptographic details involved in meeting the pubkey script requirements, typically using a suitable private key.)

In a complementary manner, each input of a transaction may be thought of as a record with three fields. The first, called an *outpoint*, is the most important for our purposes: It is composed of a transaction identifier along with the index of the desired output of the referenced transaction, in the ordered list of outputs. The second field is a *signature script* that is used to satisfy the requirements for using the referenced transaction output, as specified in the latter's *pubkey script*. (We skip these cryptographic details.) The third field is a *sequence number* that is currently unused except as part of a *locktime* feature that we ignore for simplicity.

3.2.4 Flow of Currency

Thus, Bitcoin (or satoshi) value, which is as the reward for mining blocks and awarded to an output specified by (and typically controlled by) the miner, subsequently flows through a network of transactions, with each transaction distributing all of its input value over its outputs (and transaction fee), consuming the outputs of one or more previous transactions, and generating new (unspent) outputs for potential use in subsequent transactions. These *unspent transaction outputs or UTXOs* are where satoshis are held in the system. Figure 3.7 illustrates a small portion of this *transaction graph*, i.e., the directed acyclic graph (DAG) generated by this flow of satoshis from transaction to transaction along the blockchain [46].

It is important to note that every transaction completely drains all the currency from all of its inputs. There is no *direct* way to use only part of the value in any input. However, in order to *effectively* spend only a portion of the value of an input, one may create a transaction in which part of the value is transferred to an output that is controlled (owned) by the controller of that input, so that the portion of the input that is, intuitively, not to be spent in this transaction is returned to the input's owner by way of a new output. These kinds of *change-making* transactions (or parts of transactions) provide information that is useful for clustering because one may reasonably assume a connection between the input and change-making output in such cases.

Since a transaction is meant to distribute the aggregate input value among the outputs (as specified by the value associated with each output), a constraint is that the sum of the input values must be no less than the sum of the output values. Any excess of the input values over the output values is deemed a transaction fee and is awarded to the miner of the block in which this transaction occurs. These *transaction fees* are meant to provide block miners incentives to include the paying transaction

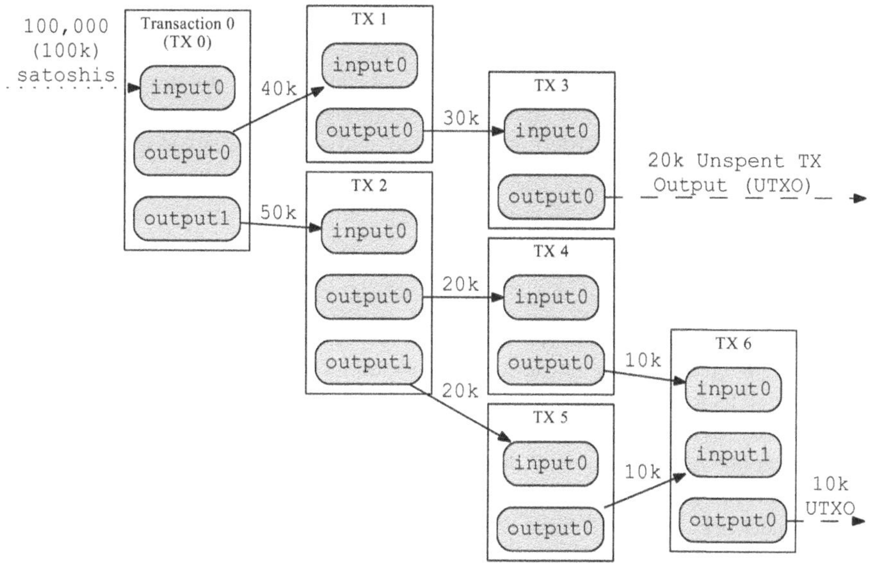

Fig. 3.7 The directed acyclic graph (DAG) formed by the flow of *satoshis* (10^{-8} Bitcoins) in the blockchain. (Figure adapted from the bitcoin.org source files, under the MIT License). The resulting *transaction graph* has transactions as vertices and a directed edge from transaction t_1 to transaction t_2 iff t_2 has an input that spends (refers to) an output of t_1. The arrows in the figure denote flow of satoshis and so point from an output in one transaction to the (at most one) input in a subsequent transaction that uses (spends) that output's value. Arrows denoting reference would point in the opposite direction, since inputs refer to prior outputs. An output is called an unspent transaction output (UTXO) in the duration between its appearance in a transaction and its spending in another (whose input points to it). Each transaction in this example includes a transaction fee of 10 k satoshi (the difference between the aggregate values of the transaction's inputs and outputs)

in the blocks they mine, sooner than other transactions that may pay less. These fees, which were historically very low (even 0), have witnessed sharp increases recently and are the subject of controversy related to the viability of certain use cases for Bitcoin. (A recent high point for average transaction fees was over 50 USD, in December 2017.) For our purposes, these transaction fees are important because they constitute a likely important attribute of transactions and, indirectly, their inputs and, through them, the controlling entities. For instance, it is likely that a person or entity has specific preferences and constraints for transaction fees. One person may prioritize speed of confirmation in the blockchain and routinely pay higher fees, while another may prefer to save on fees even if that delays a transaction. Thus the transaction fee may reveal potentially common entities across transactions even if the transactions have no other apparent connection.

The above description omits several details, notably cryptographic ones, that are not important for our purposes. For instance, verifying that the requirements specified by a pk-script are met by the signature-script of a referencing transaction is most commonly done using the *P2PKH* (pay to public key hash) scheme, which in turn uses public/private-key pairs and the *sec256k1* ECDSA (elliptic curve digital signature algorithm). There are also many other interesting aspects of the Bitcoin blockchain, such as the process of mining blocks for Bitcoin rewards, and the distributed protocol used for maintaining consensus. However, those details are unlikely to be useful for clustering and related analyses and so are skipped here.

3.3 Models of Blockchain Data

The same base blockchain dataset may be modeled and abstracted in different ways for the purposes of clustering. For example, Bitcoin transaction data may be modeled as a graph with transactions as vertices and input-to-output references as directed (reverse) edges, but also as a graph with Bitcoin addresses as vertices and payments as edges. As well, models may include the other components of the blockchain infrastructure, such as the blocks themselves and the hosts in the peer-to-peer network.

3.3.1 Transactions

A graph of blockchain transactions, similar to the DAG suggested by Fig. 3.7 is a natural model. This graph has a vertex for each transaction and a directed edge from one vertex to another if there is a transfer of satoshi (based on output–input references) between the corresponding transactions. A transaction may have multiple inputs that refer to the outputs of a common prior transaction. As a result, there may be multiple edges between the same pair of vertices (i.e., parallel edges) and the graph is in fact a multigraph. An alternative is to disallow parallel edges and to model multiple output–input transfers between a pair of transactions by labeling the edge connecting them with the number of transfers.

In addition to this primary structure, it is also important to model the secondary structure illustrated in Fig. 3.7, viz. the information on which output of a transaction is the source of satoshi for each input of another. Since the inputs of a transaction do not have any identifiers and serve only to identify outputs of prior transactions, this information may be modeled indirectly by labeling the vertices of the graph with a set of output-identifying attributes.

In more detail, each vertex is labeled with the following attributes corresponding to fields of the underlying blockchain data structures.

- One or more inputs, each with:

  previous-output: a conceptual pointer to the output of some earlier transaction that is being spent by this transaction. This pointer is encoded in an *outpoint* data type and is composed of a 32-byte transaction identifier (TXID) and a 4-byte unsigned integer that is the index of the output in the referenced transaction (see output-index below).

script-bytes: the length signature-script field that follows, with a maximum value of 10,000. This length is encoded as in a variable-length format (1 through 5 bytes) called *compactSize* that favors smaller values. For example, all values less than 253 are encoded as a single byte.

signature-script: a variable-length character array that encodes a program (script) in the Bitcoin script language. This script is used to verify that the transaction has the authority to spend the satoshis in outpoint.

sequence-number: a 4-byte integer that is currently unused (in Bitcoin Core, with a few exceptions).

- One or more outputs, each with:

output-index: the sequential number (0-based) of this output among the outputs of the transaction. This index is not stored explicitly but is implicitly determined by the order in which a transaction lists its outputs. It is used by later transactions to identify a particular output (as previous-output above).

value: the number of satoshis to send via this output, encoded as an 8-byte integer. The sum of these values for all outputs of any transaction is at most the sum of values of the outputs of previous transactions identified by the previous-output fields of the inputs of that transaction. (Recall that we ignore coinbase transactions.) Typically the former sum is strictly less than the latter, with the difference being the transaction fee awarded to the block miner.

pk-script-bytes: the length, in bytes, of the public-key script that follows. The length is encoded in the same compactSize format as used by script-bytes for inputs.

pk-script: a script in the Bitcoin script language that specifies the requirements that must be met by a future transaction in order to successfully spend this output. That transaction must meet these requirements by including the necessary details in the signature-script of an input that refers to this output.

A simple model may ignore some of the lower-level implementation details noted above. However, some of those details may provide important identification of classifying information and so it is worthwhile to include them. For example, Bitcoin implementations have some freedom in the design of the signature-script that is used to verify spending authority. An implementation or its user may intentionally or inadvertently provide additional information in such a design. A similar observation also applies to the pk-scripts used in outputs. Further, there may also be patterns in implementation details such as the order in which a Bitcoin client inserts inputs (more precisely, UTXO references) in its submitted transactions (Figs. 3.8 and 3.9).

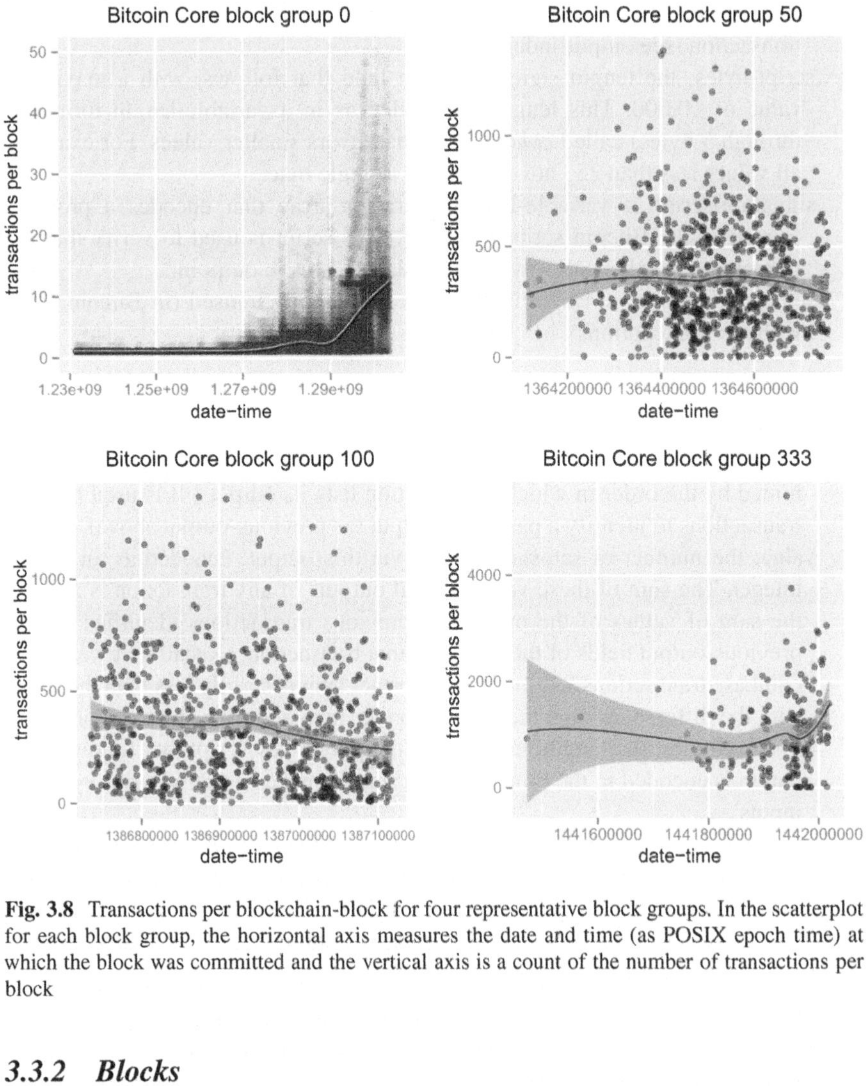

Fig. 3.8 Transactions per blockchain-block for four representative block groups. In the scatterplot for each block group, the horizontal axis measures the date and time (as POSIX epoch time) at which the block was committed and the vertical axis is a count of the number of transactions per block

3.3.2 *Blocks*

While the transaction-based model reflects an application-oriented view of blockchain data, that data may also be modeled more directly, based on the chain of blocks themselves. Compared to the transaction-based model, the model of blocks is farther removed from persons and other entities using the Bitcoin network. However, it is useful for two reasons: First, some of the information revealed at this lower level of abstraction can guide clustering and other analysis of the higher-level model. A simple example is the block timestamp and its relation to transaction timestamps. Second, some of the lower-level features may be of primary interest

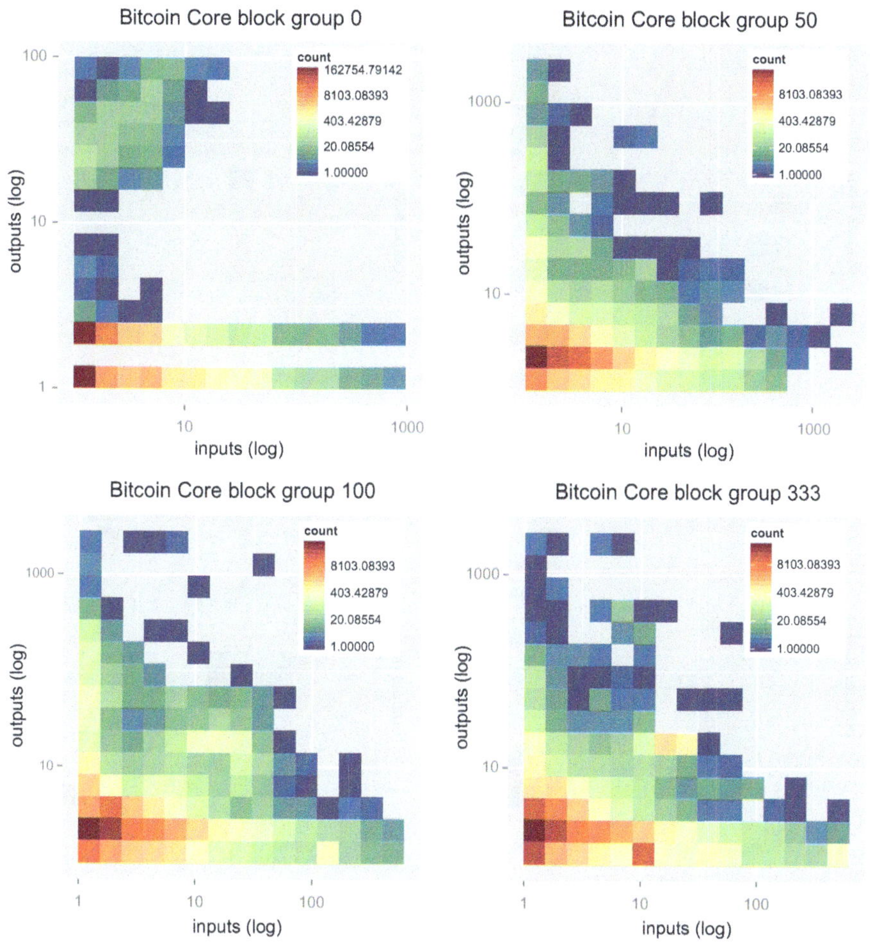

Fig. 3.9 Distribution of transactions by their input- and output-counts for four representative block groups. In the heat-map for each block group, the horizontal and vertical axes measure, using logarithmic scales, the number of inputs and (respectively) outputs of a transaction in that block group. The color of each cell is a measure of the number of transactions with the corresponding input- and output-counts, also using a logarithmic scale. Note that the scales differ significantly across the heat-maps

when the objective of clustering is gaining insights into the use and performance of the distributed system (Fig. 3.10).

Recall the simplified version of the blockchain illustrated by Fig. 3.5. In more detail, each block includes the following fields:

version: a 4-byte integer that determines the validation rules applicable to the block.

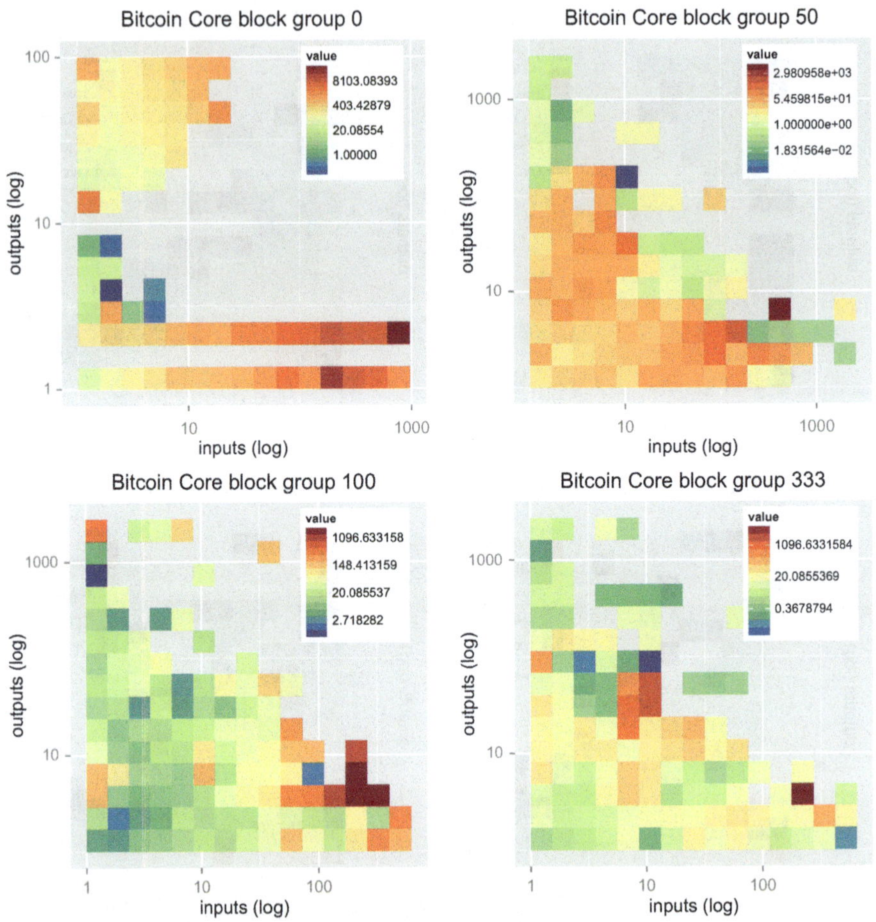

Fig. 3.10 Distribution of transacted Bitcoin value (XBT) over transaction input- and output-counts for four representative block groups. In the heat-map for each block group, the horizontal and vertical axes measure, using logarithmic scales, the number of inputs and (respectively) outputs of a transaction in that block group (as in Fig. 3.9). The color of each cell is a measure of the total Bitcoin value transferred in transactions with the corresponding input- and output-counts, also using a logarithmic scale. Note that the scales differ significantly across the heat-maps. Recall, from Fig. 3.1, that the exchange rate for XBT relative to fiat currencies such as USD varies greatly with time

previous-block-header-hash: a 32-byte double-SHA256 hash that serves as a pointer to the previous block in the chain (in addition to its cryptographic role).

merkle-root-hash: a 32-byte double-SHA256 hash of the root of the transactions' Merkle tree.

time: a 4-byte unsigned integer encoding the time when this block's miner started hashing the header (self-reported by miner). The value is interpreted as

POSIX epoch time, roughly (modulo leap seconds) the number of seconds since midnight (UTC) on 1970-01-01.

target/n-bits: the maximum numerical value permitted for the hash of this block's header. This target determines difficulty of mining a block. It is a 256-bit unsigned integer that is encoded less precisely, using only 32 bits, in the n-bits field.

nonce: a 4-byte number that is inserted by the block's miner in order to meet the hash-target requirement above without changing other block data.

block-height: distance of the block (along the chain) from the genesis block (very first block mined).

coinbase transaction: the very first transaction encoded in each block is a special one that collects and spends the *block reward* (block subsidy plus transaction fees) at the discretion of the block miner (typically paid to an address owned by the miner).

other transactions: a list of transactions, each with several attributes, described in Sect. 3.3.1.

Some of the above attributes concern details of the blockchain protocol and may, at first glance, appear irrelevant to clustering or other similar analyses. However, some details, such as the version, will be relevant for long-term analysis that spans versions. As well, the coinbase transaction, in addition to the output address, includes an input field that is unconstrained (unlike those for other transactions, described below). The contents of this field may be used by the block's miner to encode useful data, either purposefully or as a consequence of the specific software used. In the Bitcoin genesis block [6], this field contains the string "The Times 03/Jan/2009 Chancellor on brink of second bailout for banks" (a front-page headline from the Financial Times on that date, presumably included as a proof of recency of the genesis block) [8].

3.3.3 Addresses

While the transaction graph above provides a faithful representation of the information encoded in the blockchain, it is often useful to raise the level of abstraction from transactions, and their inputs and outputs, to transfers among Bitcoin addresses. In such an *address graph* each vertex represents a Bitcoin address. A directed edge (u, v) labeled w denotes the transfer of w satoshis from the address corresponding to u to that for v. Unlike transaction outputs, addresses may, and often are, reused, so there are in general multiple out-edges from and in-edges to a vertex in this graph. Further, since there may be multiple transfers between the same ordered pair of addresses, as well as transfers from an address to itself (e.g., in a change-making transaction), the resulting structure is a multigraph with, typically, parallel edges and self-loops.

3.3.4 Owners

It is convenient to assume that each Bitcoin address is owned and controlled by at most one entity (person, organization, etc.), called its *owner*. (At most one instead of exactly one because addresses may become ownerless if credentials are somehow lost.) Strictly speaking, this assumption is not valid because it is possible for multiple people to share an address (akin to sharing of email accounts). However, the use cases for such sharing are much fewer than those for individually owned addresses and this assumption is commonly made by prior work.

More concretely, we may describe the model as a directed multigraph with vertices representing owners and a directed edges representing transfer of satoshi. This graph is very similar to the address graph and, like that graph, is more precisely a multigraph with (likely) parallel edges (representing multiple transfers between the same pair of owners) and self-loops (representing transfers made by change-making and similar transactions).

Since there is no systemic association of Bitcoin addresses with owners, modeling the blockchain in this manner requires indirect approaches. Indeed, such a model is often the desired result of clustering or other analyses.

3.3.5 Nodes

In addition to modeling the above conceptual components of the blockchain, we may also model the blockchain at a lower level of abstraction by focusing on the *nodes*, i.e., the peer-to-peer computer hosts that maintain the distributed blockchain. Doing so provides opportunities for clustering and inference not otherwise available [29, 32, 53]. It may be of particular interest in private or limited-access blockchains.

Each node may be modeled by:

IP-address. A node's IP network address may be used as its identifier, ignoring complications such as multi-homed nodes.

Initiated-transactions. Each transaction may be conceptually assigned to an initiating node based on the times at which the transaction is publicized by the nodes in the network.

Volume. Related to the above, the volume of a node is the total value of Bitcoins initiated at that node.

Time-series. Instead of using only snapshot or aggregate attributes of a node, the variation for that attribute (e.g., total daily value of initiated transactions) can be modeled as a time series.

In addition to their other attributes, nodes may be clustered by the transactions they initiate. In turn, clustered nodes may be used to cluster transactions based on their originating nodes.

3.4 Clustering

At a high level, there are two approaches for clustering blockchain data. The first approach is based on applying mostly conventional, well- studied clustering algorithms (e.g., k-means) and robust implementations to feature-vectors extracted from blockchain datasets [25, 30, 33, 62]. The key tasks in this case are determining the set of features to use, extracting them effectively from the underlying data, and transforming them (e.g., scaling, normalization) to improve clustering. Section 3.4.1 describes feature extraction for this purpose. The associated challenge of scaling clustering methods to the size and throughput of blockchain datasets is an important one as well, but one that may be addressed using well-developed prior work on clustering large datasets in general.

The second approach is based on a more direct utilization of the particular characteristics of blockchain data (e.g., co-occurrence of transaction inputs, or TXIs, or other temporal patterns [19]) and their associated semantics (e.g., identical or closely related owners). The key tasks in this case are determining which semantics, assumptions, and heuristics to use for the purpose of forming clusters, and designing algorithms tailored to those that are capable of efficiently operating on the large datasets involved. Section 3.4.2 describes such options in the context of the address graph (Sect. 3.3.3), where the focus is on determining which vertices (addresses) are likely to represent the same off-network user or other entity, i.e., on correlating the address graph with the owner graph (Sect. 3.3.4).

3.4.1 Feature Extraction

Prior work has identified and used several features from the underlying data for clustering and other purposes. For instance, the features extracted from transaction data (Sect. 3.3.1) and used by recent work on fraud detection based on trimmed k-means clustering [41] are categorized as:

currency values: the sum, mean, and standard-deviations of the amounts received and sent.

network properties: in- and out-degrees, clustering coefficient (described below), and number of triangles.

neighborhood: based on the sets of transactions from which a given transaction receives (and sends) satoshi, i.e., its in-neighborhood (and, respectively, out-neighborhood).

Intuitively, the local *clustering coefficient* [56] for a vertex measures the degree of interconnectedness among its neighbors. In more detail, it measures how similar the subgraph induced by these neighbors is to a fully connected (complete) graph on those vertices. Formally, the local clustering coefficient $\mathrm{cc}(v)$ of a vertex v in a directed graph D is defined as:

$$\mathrm{cc}(v) = \frac{|(N_D^+(v) \times N_D^+(v)) \cap E(D)|}{d_D^+(v)(d_D^+(v) - 1)}$$

where, following standard notation [9], we use $V(D)$ and $E(D)$ to denote the vertices and edges of a directed graph D, and where

$$N_D^+(v) = \{u \in V(D) \mid (v, u) \in E(D)\}$$

is the out-neighborhood of v and $d_D^+(v) = |N_D^+(v)|$ is the out-degree of v.

These extracted features were used as inputs to the classic and trimmed versions of the k-means clustering algorithm [15, 17, 47]. A curve of within-group distances versus number of clusters suggested $k = 8$ as a suitable parameter for the data in that study.

In similar work on anomaly detection [49], three features were used for the transactions graph (similar to the one of Sect. 3.3.1): in-degree, out-degree, and total value. The same study used six other features for the address graph (similar to the one of Sect. 3.3.3) were in- and out-degrees, means of incoming and outgoing transaction values, mean time interval, and clustering coefficient. In order to account for the widely varying scales and large ranges of the feature metrics, the normalized logarithm of their values is used for further processing.

3.4.2 Address Merging

In principle, it is possible (and often recommended) that a Bitcoin address be used exactly once to receive funds and exactly once to send funds, with unused funds sent to a new address controlled by the owner of the first one. However, it is quite common for an address to be reused. A common case is the use of Bitcoin addresses embedded into software user interfaces or in Web pages to solicit donations. These kinds of addresses are long lived and participate in several transactions. Many such reused addresses also explicitly identify their owners, such as a software development team or a person. (As well, much like vanity license plates for vehicles, there are vanity addresses that include the owner's name, or desired tag, as a substring when expressed in the base-58 encoding used by Bitcoin [57].) Such identified addresses are important resources not only because of the direct identification they provide but also because they may be used to bootstrap methods that can identify or classify addresses that are otherwise anonymous.

In this context, an important task that has been the focus of prior work is *address merging*, i.e., finding sets of Bitcoin addresses that are likely to be owned and controlled by a common person or entity. Some methods for merging Bitcoin addresses are below [16, 38, 53]:

Co-occurring transaction inputs. All the addresses that are referenced by inputs of a common transaction are assumed to have a common controlling owner. We may recall that transaction inputs (TXIs) do not include any addresses. However, these addresses are available in the outputs (TXOs) of prior transactions to which these inputs refer by providing transaction identifiers and output indices [60].

Transaction input–output patterns. Although the intent of transactions is not explicitly available, in some cases it may be reasonably inferred based on patterns. These patterns may use the number and variety of transaction inputs and outputs and, importantly, the incoming and outgoing distributions of value (satoshi). For example, it is very easy to determine, with high confidence, which transactions are payments to members of a Bitcoin mining pool, based on the total value (equal or close to the current block reward), and the input–output distribution: a single or very few inputs, representing the mining pool, and a large number of outputs, representing the pool members.

Another important class of transactions in this context is the class of change-making transactions. Recall that there is no direct way to use only a part of the value available in an unspent transaction output (UTXO). Rather, that effect must be achieved by spending the entire UTXO value in a transaction, but paying part of it to an output that is controlled by the owner of the source UTXO. Since change-making is an essential operation for many real-world transactions, the corresponding blockchain transactions are also frequent. Again, there is no explicit marker for such transactions. However, they may be detected by using a variety of techniques.

One effective technique is based on the observed practice (though not requirement) of Bitcoin wallet programs generating completely new addresses on behalf of their users whenever change-making is required [38]. The resulting observable pattern in a transaction is an address A with the following properties:

- The transaction is not a coinbase (mining reward claiming) transaction.
- There is no prior transaction that includes A in its output. (It follows that A cannot be referenced by any transaction's input as well.)
- Among all the outputs of this transaction, A is the only one satisfying the above conditions.

Peer host address. While the above methods rely on information from the transaction model of blockchain data, some information from other models, such as the node model, may also be used to aid transaction-based clustering. An obvious candidate is the IP address of the peer host that from that originates a transaction. There is no explicit field in transactions that records its originating peer. At first glance, there is no way to determine whether a transaction arriving form a peer is one originated at that peer or forwarded by the peer on behalf of another, recursively. However, by using a client that simultaneously connects to a large number of peers, it is possible to use timing information to make inferences about the likely origins of a transaction. That is, if a client is connected to a large number of peers, then it is reasonable to assume that the peer from which a transaction was first received is very likely the peer at which it originated.

Temporal patterns. When transactions matching various patterns (as above) are clustered and tagged, the clusters may be refined by observing the temporal patterns of the transactions. For example, if a cluster contains 200 transactions, with 100 of them occurring at roughly daily intervals, 70 at weekly, and 30 at monthly, that cluster may be subdivided accordingly.

Well-known services. Certain well-known services have transaction patterns that allow association of input and output addresses. For example, the *Satoshi Dice* service has been studied in this context and used to connect addresses [38]. Well-known services may also be used to detect when addresses merged by other rules are likely separately controlled and should therefore be separated. For this purpose, recent work [20] divides Bitcoin services into six categories and defined a pairwise compatibility matrix for those categories. For example, gambling services are deemed incompatible with pool services, while they are compatible with exchanges. Addresses that appear in incompatible pairs of categories are deemed erroneously merged and so separated.

3.4.3 Scalability

Our focus has been on understanding blockchain data and on framing the space of clustering as applied to that data. Compared to other clustering's typical application domains, work on clustering applied to blockchains is still at a very early stage. So it is appropriate to focus on problem definitions and output-quality evaluation metrics over performance and scalability issues such as running time and space requirements. Nevertheless, it is appropriate to briefly consider such scalability issues here.

General-purpose density-based clustering algorithms, such as the many variants of the DBSCAN algorithm, are good candidates because they are not only well studied but also implemented widely in several popular libraries and systems [1, 26, 36]. Specialized algorithms for outlier detection and network clustering are also applicable [54, 59, 63]. The performance of DBSCAN and related algorithms strongly depends on the nature of the data, the critical ϵ and *MinPts* parameters, and on the availability of suitable indexes [24, 55]. Given the volume of blockchain data, parallel execution of clustering algorithms is a useful strategy. For example, recent work has demonstrated a DBSCAN-like algorithm that is designed for the MapReduce framework [28].

3.5 Evaluation

The most direct methods for evaluating the results of a clustering algorithm, and related indices such as *purity* and *entropy* [37], are based on determining how well the algorithm detects clusters that have been determined by some other method, typically with human input [18]. For blockchain data, such an *external* evaluation method is of very limited effectiveness due to the scarcity of such tagged and well-studied datasets. The volume of data, as well as other characteristics, mean that it is impracticable to rely on the availability of such test data in the near future as well. It is therefore necessary to use methods that do not rely on human-studied datasets but, instead, use some other intrinsic characteristics of the input data and output clusters, i.e., an *internal* method [58].

3.5.1 Distance-Based Criteria

Such intrinsic criteria for evaluating clusters have been studied in the domain of document clustering in the vector-space model [51]. That work builds on earlier work [18] and on the intuitive desirability of *compactness* and *isolation* of clusters.

3.5.1.1 Cluster Quality Criteria

In particular, four quality criteria and associated indices are presented. (The following presentation uses different, more descriptive, names for the criteria and modified, but equivalent, formal definitions of the indices.)

Total intra- and inter-cluster distance. This criterion seeks to minimize both the separation of objects that are in the same cluster and the separation of clusters. More precisely, let C denote a collection of clusters, d a distance metric for the elements (vectors) being clustered, and $\odot(C)$ the centroid of cluster C:

$$\odot(C) = \frac{1}{|C|} \sum_{\mathbf{x} \in C} \mathbf{x} \tag{3.1}$$

The *dst* index that this criterion seeks to minimize may then be expressed as follows:

$$\mathrm{dst}(C) = \sum_{C \in C} \sum_{x \in C} d(x, \odot(C)) + \sum_{C \in C} d(\odot(C), \odot(\cup C)) \tag{3.2}$$

The first term on the right-hand side quantifies intra-cluster distances, as it sums the distances of elements from the centroids of their assigned clusters. The second quantifies inter-cluster distances, as it sums the distances of cluster centroids from the global centroid. Minimizing the intra-cluster distances is a natural expression of the general preference for denser clusters. In contrast, minimizing the inter-cluster distances may appear counterintuitive at first because it contradicts the general preference for well-separated clusters. However, the intuitive effect of this criterion is to strike a balance between the desire for compact clusters (which, on its own, could be satisfied by putting each object in its own cluster) and the desire for fewer clusters (which, on its own, could be satisfied by putting all objects in a single cluster. This distance has the same value in both these extreme cases, and is lower in intermediate cases.

Cluster separation. Intuitively, this criterion favors compact, well-separated clusters by comparing element-wise within-cluster distances with distances between cluster centroids. More precisely, for a given cluster C, we may compute the ratio of the maximum of the distances between pairs of elements in that cluster and the minimum of the distances from the centroid of C to other clusters. The *eld* index is defined as the reciprocal of the sum of this ratio over all clusters that contain at least two elements. This index is undefined for the two extreme cases of one cluster per object and one cluster for all objects.

$$\text{sep}(\mathcal{C}) = \left(\sum_{\substack{C \in \mathcal{C} \\ |C|>1}} \frac{\max\{d(x_i, x_j) \mid x_i, x_j \in C\}}{\min\{d(\odot(C), \odot(C')) \mid C \neq C' \in \mathcal{C}\}} \right)^{-1} \tag{3.3}$$

Element-wise distances. The intuition behind this criterion is that, in good clusterings, the largest distance between any pair of elements in a cluster should be small, and the smallest distance between a pair of elements in different clusters should be large. More precisely, the *eld* index (which is to be minimized) sums, for each element, the difference between its distances to the farthest object in its cluster and the nearest object in some other cluster. This index has its maximum value when there is a single cluster with all elements. Its value is also large when elements are in singleton clusters. It favors intermediate clusterings that have large clusters.

$$\text{eld}(\mathcal{C}) = \sum_{C \in \mathcal{C}} \sum_{x \in C} \left(\max_{y \in C} d(x, y) - \min_{z \notin C} d(x, z) \right) \tag{3.4}$$

Intra- and inter-cluster similarity. Unlike the earlier criteria, this criterion depends on a method for evaluating the similarity S of two elements, and quantifies the intuition that intra-cluster similarities should be large while inter-cluster similarities should be small. More precisely, the *sim* index (which is to be maximized) sums, over all clusters, the difference between the total pairwise similarity for elements in the cluster and the total pairwise similarities with one object in and one object out of the cluster. This index favors large clusters over smaller clusters. Indeed, in the extreme case of one cluster per object, its value is negative, while in the other extreme of all objects in a single cluster, its value is maximized.

$$\text{sim}(\mathcal{C}) = \sum_{C \in \mathcal{C}} \left(\sum_{x,y \in C} S(x, y) - \sum_{\substack{x \in C \\ z \notin C}} S(x, z) \right) \tag{3.5}$$

3.5.1.2 Mahalanobis Distance

As a variation on the general theme of the above criteria, it may be useful to measure the distance between an element and its assigned cluster using a metric such as the Mahalanobis distance in order to better account for skewed cluster shapes and diverse variances along different dimensions [35].

If V is a set of elements (vectors) with mean $\boldsymbol{\mu}$ covariance matrix $S = (\sigma_{ij})$, then the Mahalanobis distance of an element $\mathbf{v}$ from E is

$$d_m(\mathbf{v}, V) = \sqrt{(\mathbf{v} - \boldsymbol{\mu})^T S^{-1} (\mathbf{v} - \boldsymbol{\mu})} \tag{3.6}$$

For instance, the two terms inside the parentheses in Eq. (3.5) may be replaced by the Mahalanobis distance of x from the set of points in (respectively, not in) the x's cluster.

Work using the Mahalanobis distance for outlier detection in the address- and transaction-graphs found a strong correlation between the detected outliers and the boundaries of scatterplots of in- and out-degrees of vertices [49]. This result appears unsurprising in this case because normalized values of the same features are used for determining outliers. However, it does indicate that these features likely have the dominant influence for that method.

3.5.2 *Sensitivity to Cluster Count*

With other parameters held constant, as we increase the number of clusters (especially in a method such as k-means), we expect the distance of each element from the centroid of its assigned cluster to decrease. However, the rate at which the latter distance decreases will, in general, vary across clustering methods. It is natural to prefer methods that provide a more rapid reduction in this distance as the number of clusters increases. This intuition can be made more concrete by plotting a curve with number-of-clusters on the horizontal axis and average distance of elements from the cluster centroids on the vertical axis. The desired criterion then maps to seeking curves with a small area between the curve and the horizontal axis. If we invert the vertical axis, then we seek to maximize the area under the curve. In this respect, this criterion is similar to that used in the receiver-operator characteristic (ROC) [21] for studying the trade-off between true positives and false positives. While there is graphical similarity, the underlying concepts may be different in some cases. However, when clustering is being used for a classification task such as fraud detection, the concepts are similar. For instance, such a curve is used in some work that evaluates clustering for Bitcoin fraud detection [41].

3.5.3 Tagged Data

There is a small amount of tagged blockchain data that can be used for evaluating clustering, outlier detection, and related methods. This data is typically derived from well-known cases of theft or other fraudulent activity on the Bitcoin network. The scarcity of such tagged data makes it unsuitable for use in a primary evaluation method; however, it is valuable as a secondary check on the effectiveness of methods in detecting confirmed thefts or other tagged activity.

While such data has been mainly gathered passively, in response to reports of Bitcoin thefts or similar events, there is also some work [38] on actively tagging a subset of blockchain data using store purchases and other activities that trigger Bitcoin transactions, and correlating the results.

There are also some services that associate user-defined tags with Bitcoin addresses [7, 31]. Since these services are easily updated by anyone without any restrictions, there are no guarantees about the accuracy of the tags and associations. Nevertheless, the information could be used in a secondary manner for clustering, for example, to assign descriptive tags to clusters identified using more reliable data.

Another related method uses synthetically tagged data, such as outliers detected by another method, to evaluate results by measuring the distance of the outliers from the nearest cluster centroids [49]. In that work, the focus is on detecting outliers using clustering (k-means) as a baseline, but the same strategy could be used in the opposite direction. A related strategy, also used in that work, is to use the results on one model of the underlying data to evaluate results on another model, using known or assumed mappings between concepts in the two models. In particular, results on the transactions-as-vertices graph model can be used to evaluate results on the owners-as-vertices model, and vice versa.

3.5.4 Human-Assisted Criteria

Although, as noted earlier, it is often not convenient or possible to analyze blockchain datasets with human feedback, some human-assisted methods can augment the automated methods. For this purpose, it is important that the clusters and their important features be available for multiple commonly used visualizations, such as scatterplots, histograms, and parallel-coordinates plots.

In some prior work, such human-assisted criteria have been profitably used to improve the clustering process and to validate some of the results. For example, a study on de-anonymizing Bitcoin addresses used a graphical visualization of the user network [38]. That work also illustrates the use of several other visualizations of the blockchain data for making inferences about the dominant modes in which currency moves through the network of user addresses. Similarly, a visual representation of the Bitcoin transactions graph has been used to detect communities by using a few labeled nodes as starting points [22].

3.6 Conclusion

Blockchains are a rich and growing source of big data. The major blockchains underlying cryptocurrencies, such as Bitcoin, provide a complete and public record of all transactions, large and small, that is unprecedented in its size and scope. In addition to Bitcoin and other cryptocurrencies such as Ethereum, Ripple, and more, there are also other, non-cryptocurrency applications of blockchains. All these developments suggest an increasing prevalence and importance of blockchain big data.

The availability of this data presents opportunities for discovering clusters and other patterns for purposes such as detecting common usage patterns and trends, financial fraud, and other activities, such as gaming, that leave a financial trail. Since cryptocurrencies are sometimes used to process payments for illegal schemes (such as cyber-ransom), de-anonymizing some transactions and addresses has received special attention.

In addition to its volume, velocity, and variety, big data from blockchains presents some additional challenges to clustering and related analyses. Many of these stem from the details of how a blockchain system encodes and processes transactions, typically pseudo-anonymously, i.e., without any explicit identification other than randomly generated addresses. Another significant challenge is the sparsity of information on datasets for study and, especially, for evaluating clustering methods. While these challenges are formidable, recent work has demonstrated several promising techniques to address them, such as using a modest amount of off-blockchain data to seed or otherwise improve the clustering of addresses and other blockchain data. Developing such methods to cluster blockchain data and, in particular, to evaluate the results of clustering is a promising avenue for further work. Another promising direction is applying clustering and other unsupervised learning techniques to the formal contracts provided by Ethereum [44].

Acknowledgements This work was supported in part by the US National Science Foundation grants EAR-1027960 and PLR-1142007. Several improvements resulted from detailed feedback from the reviewers.

References

1. M. Ankerst, M.M. Breunig, H.P. Kriegel, J. Sander, Optics: ordering points to identify the clustering structure, in: *Proceedings of the 1999 ACM SIGMOD International Conference on Management of Data, SIGMOD'99*. ACM, New York (1999), pp. 49–60. https://doi.org/10.1145/304182.304187
2. M.K. Awan, A. Cortesi, Blockchain transaction analysis using dominant sets, in *Computer Information Systems and Industrial Management*, ed. by K. Saeed, W. Homenda, R. Chaki. Springer, Cham (2017), pp. 229–239
3. L. Backstrom, C. Dwork, J. Kleinberg, Wherefore art thou R3579X? Anonymized social networks, hidden patterns, and structural steganography, in *Proceedings of the 16th International World Wide Web Conference* (2007)

4. G. Becker, Merkle signature schemes, Merkle trees and their cryptanalysis. Seminararbeit, Ruhr-Universität Bochum (2008). https://www.emsec.rub.de/media/crypto/attachments/files/2011/04/becker_1.pdf
5. Bitcoin price—time series—daily (2018). https://docs.google.com/spreadsheets/d/1cdP-AArCNUB9jS8hEYFFC1qxp4DMEpBCvvC5yuopD68/
6. Bitcoin Genesis Block, Blockchain.info Blockchain Explorer (2009). https://blockchain.info/tx/4a5e1e4baab89f3a32518a88c31bc87f618f76673e2cc77ab2127b7afdeda33b
7. Blockchain Luxembourg S.A., Address tags. Bitcoin address tags database (2018). https://blockchain.info/tags
8. Blockchain Luxembourg S.A., Blockchain explorer (2018). https://blockchain.info/
9. J. Bondy, U. Murty, *Graph Theory* (Springer, London, 2008)
10. J. Bonneau, A. Miller, J. Clark, A. Narayanan, J.A. Kroll, E.W. Felten, SoK: research perspectives and challenges for Bitcoin and cryptocurrencies, in *Proceedings of the 36th IEEE Symposium on Security and Privacy*, San Jose, California (2015), pp. 104–121
11. V. Buterin, et al., Ethereum whitepaper (2013). https://github.com/ethereum/wiki/wiki/White-Paper
12. Chainanalysis, Inc., Chainanalysis reactor (2018). https://www.chainalysis.com/
13. CoinMarketCap, Historical data for Bitcoin (2018). https://coinmarketcap.com/currencies/bitcoin/historical-data/
14. K. Collins, Inside the digital heist that terrorized the world—and only made $100k. Quartz (2017). https://qz.com/985093/inside-the-digital-heist-that-terrorized-the-world-and-made-less-than-100k/
15. J.A. Cuesta-Albertos, A. Gordaliza, C. Matran, Trimmed k-means: an attempt to robustify quantizers. Ann. Stat. **25**(2), 553–576 (1997)
16. D. Di Francesco Maesa, A. Marino, L. Ricci, Uncovering the Bitcoin blockchain: an analysis of the full users graph, in *2016 IEEE International Conference on Data Science and Advanced Analytics (DSAA)* (2016), pp. 537–546. https://doi.org/10.1109/DSAA.2016.52
17. C. Ding, X. He, K-means clustering via principal component analysis, in *Proceedings of the Twenty-first International Conference on Machine Learning, ICML'04* (ACM, Banff, 2004), p. 29. https://doi.org/10.1145/1015330.1015408
18. R. Dubes, A.K. Jain, Validity studies in clustering methodologies. Pattern Recogn. **11**, 235–254 (1979)
19. A. Epishkina, S. Zapechnikov, Discovering and clustering hidden time patterns in blockchain ledger, in *First International Early Research Career Enhancement School on Biologically Inspired Cognitive Architectures* (2017)
20. D. Ermilov, M. Panov, Y. Yanovich, Automatic Bitcoin address clustering, in *Proceedings of the 16th IEEE International Conference on Machine Learning and Applications (ICMLA)*, Cancun, Mexico (2017)
21. T. Fawcett, ROC graphs: notes and practical considerations for researchers. Pattern Recogn. Lett. **27**(8), 882–891 (2004)
22. M. Fleder, M.S. Kester, S. Pillai, Bitcoin transaction graph analysis. CoRR (2015). abs/1502.01657
23. B. Fung, Bitcoin got a big boost in 2017. Here are 5 other cryptocurrencies to watch in 2018. Washington Post—Blogs (2018)
24. J. Gan, Y. Tao, Dbscan revisited: mis-claim, un-fixability, and approximation, in *Proceedings of the 2015 ACM SIGMOD International Conference on Management of Data, SIGMOD'15* (ACM, New York, 2015), pp. 519–530. https://doi.org/10.1145/2723372.2737792
25. Z. Ghahramani, Unsupervised learning, in *Advanced Lectures on Machine Learning*, ed. by O. Bousquet, U. von Luxburg, G. Rätsch. Lecture Notes in Computer Science, vol. 3176, chap. 5 (Springer, Berlin, 2004), pp. 72–112
26. A. Gunawan, A faster algorithm for DBSCAN. Master's Thesis, Technical University of Eindhoven (2013)
27. M. Harrigan, C. Fretter, The unreasonable effectiveness of address clustering, in *International IEEE Conferences on Ubiquitous Intelligence and Computing, Advanced and Trusted Com-*

puting, Scalable Computing and Communications, Cloud and Big Data Computing, Internet of People, and Smart World Congress (UIC/ATC/ScalCom/CBDCom/IoP/SmartWorld) (2016), pp. 368–373. https://doi.org/10.1109/UIC-ATC-ScalCom-CBDCom-IoP-SmartWorld.2016.0071
28. Y. He, H. Tan, W. Luo, S. Feng, J. Fan, MR-DBSCAN: a scalable MapReduce-based DBSCAN algorithm for heavily skewed data. Front. Comp. Sci. **8**(1), 83–99 (2014)
29. B. Huang, Z. Liu, J. Chen, A. Liu, Q. Liu, Q. He, Behavior pattern clustering in blockchain networks. Multimed. Tools Appl. **76**(19), 20099–20110 (2017). https://doi.org/10.1007/s11042-017-4396-4
30. A.K. Jain, M.N. Murty, P.J. Flynn, Data clustering: a review. ACM Comput. Surv. **31**(3), 264–323 (1999). https://doi.org/10.1145/331499.331504.
31. A. Janda, WalletExplorer.com: smart Bitcoin block explorer (2018). Bitcoin block explorer with address grouping and wallet labeling
32. D. Kaminsky, Black ops of TCP/IPi. Presentation slides (2011). http://dankaminsky.com/2011/08/05/bo2k11/
33. T. Kohonen, Essentials of the self-organizing map. Neural Netw. **37**, 52–65 (2013). https://doi.org/10.1016/j.neunet.2012.09.018. Twenty-fifth Anniversary Commemorative Issue
34. H. Kuzuno, C. Karam, Blockchain explorer: an analytical process and investigation environment for Bitcoin, in *Proceedings of the APWG Symposium on Electronic Crime Research (eCrime)* (2017), pp. 9–16. https://doi.org/10.1109/ECRIME.2017.7945049
35. P.C. Mahalanobis, On the generalised distance in statistics. Proc. Natl. Inst. Sci. India **2**(1), 49–55 (1936)
36. S.T. Mai, I. Assent, M. Storgaard, AnyDBC: an efficient anytime density-based clustering algorithm for very large complex datasets, in *Proceedings of the 22nd ACM SIGKDD International Conference on Knowledge Discovery and Data Mining*, KDD'16 (ACM, New York, 2016), pp. 1025–1034. https://doi.org/10.1145/2939672.2939750
37. J. McCaffrey, Data clustering using entropy minimization. Visual Studio Magazine (2013)
38. S. Meiklejohn, M. Pomarole, G. Jordan, K. Levchenko, D. McCoy, G.M. Voelker, S. Savage, A fistful of Bitcoins: characterizing payments among men with no names, in *Proceedings of the Conference on Internet Measurement, IMC'13*, (ACM, Barcelona, 2013), pp. 127–140. https://doi.org/10.1145/2504730.2504747
39. R.C. Merkle, A digital signature based on a conventional encryption function, in *Advances in Cryptology—CRYPTO'87*, ed. by C. Pomerance (Springer, Berlin, 1988), pp. 369–378
40. I. Miers, C. Garman, M. Green, A.D. Rubin, Zerocoin: anonymous distributed e-cash from Bitcoin, in *Proceedings of the IEEE Symposium on Security and Privacy* (2013)
41. P. Monamo, V. Marivate, B. Twala, Unsupervised learning for robust Bitcoin fraud detection, in *Proceedings of the 2016 Information Security for South Africa (ISSA 2016) Conference*, Johannesburg, South Africa (2016), pp. 129–134
42. C.M. Nachiappan, P. Pattanayak, S. Verma, V. Kalyanaraman, Blockchain technology: beyond Bitcoin. Technical Report, Sutardja Center for Entrepreneurship & Technology, University of California, Berkeley (2015)
43. S. Nakamoto, Bitcoin: a peer-to-peer electronic cash system. Pseudonymous posting (2008). Archived at https://bitcoin.org/en/bitcoin-paper
44. R. Norvill, B.B.F. Pontiveros, R. State, I. Awan, A. Cullen, Automated labeling of unknown contracts in ethereum, in *Proceedings of the 26th International Conference on Computer Communication and Networks (ICCCN)*, (2017), pp. 1–6. https://doi.org/10.1109/ICCCN.2017.8038513
45. M. Ober, S. Katzenbeisser, K. Hamacher, Structure and anonymity of the Bitcoin transaction graph. Future Internet **5**(2), 237–250 (2013). https://doi.org/10.3390/fi5020237. http://www.mdpi.com/1999-5903/5/2/237
46. M.S. Ortega, The Bitcoin transaction graph—anonymity. Master's Thesis, Universitat Oberta de Catalunya, Barcelona (2013)
47. V.C. Osamor, E.F. Adebiyi, J.O. Oyelade, S. Doumbia, Reducing the time requirement of k-means algorithm. PLoS One **7**(12), 1–10 (2012). https://doi.org/10.1371/journal.pone.0049946

48. S. Patel, Blockchains for publicizing available scientific datasets. Master's Thesis, The University of Mississippi (2017)
49. T. Pham, S. Lee, Anomaly detection in Bitcoin network using unsupervised learning methods (2017). arXiv:1611.03941v1 [cs.LG] https://arxiv.org/abs/1611.03941v1
50. S. Pongnumkul, C. Siripanpornchana, S. Thajchayapong, Performance analysis of private blockchain platforms in varying workloads, in *Proceedings of the 26th International Conference on Computer Communication and Networks (ICCCN)* (2017), pp. 1–6. https://doi.org/10.1109/ICCCN.2017.8038517
51. B. Raskutti, C. Leckie, An evaluation of criteria for measuring the quality of clusters. in *Proceedings of the 16th International Joint Conference on Artificial Intelligence—Volume 2, IJCAI'99*. Stockholm, Sweden (1999), pp. 905–910. http://dl.acm.org/citation.cfm?id=1624312.1624348
52. S. Raval, Decentralized applications: harnessing Bitcoin's blockchain technology. O'Reilly Media (2016). ISBN-13: 978-1-4919-2454-9
53. F. Reid, M. Harrigan, An analysis of anonymity in the Bitcoin system (2012). arXiv:1107.4524v2 [physics.soc-ph]. https://arxiv.org/abs/1107.4524
54. E. Schubert, A. Koos, T. Emrich, A. Züfle, K.A. Schmid, A. Zimek, A framework for clustering uncertain data. Proc. VLDB Endow. **8**(12), 1976–1979 (2015). https://doi.org/10.14778/2824032.2824115
55. E. Schubert, J. Sander, M. Ester, H.P. Kriegel, X. Xu, DBSCAN revisited, revisited: why and how you should (still) use DBSCAN. ACM Trans. Database Syst. **42**(3), 19:1–19:21 (2017). https://doi.org/10.1145/3068335
56. D.J. Watts, S.H. Strogatz, Collective dynamics of 'small-world' networks. Nature **393**, 440–442 (1998)
57. What is Bitcoin vanity address? (2017). http://bitcoinvanitygen.com/
58. H. Xiong, J. Wu, J. Chen, K-means clustering versus validation measures: A data distribution perspective, in *Proceedings of the 12th ACM SIGKDD International Conference on Knowledge Discovery and Data Mining, KDD'06*, Philadelphia, PA, USA (2006), pp. 779–784. https://doi.org/10.1145/1150402.1150503
59. X. Xu, N. Yuruk, Z. Feng, T.A.J. Schweiger, Scan: a structural clustering algorithm for networks, in *Proceedings of the 13th ACM SIGKDD International Conference on Knowledge Discovery and Data Mining, KDD'07* (ACM, New York, 2007), pp. 824–833. https://doi.org/10.1145/1281192.1281280
60. Y. Yanovich, P. Mischenko, A. Ostrovskiy, Shared send untangling in Bitcoin. The Bitfury Group white paper (2016) (Version 1.0)
61. J. Yli-Huumo, D. Ko, S. Choi, S. Park, K. Smolander, Where is current research on blockchain technology?—a systematic review. PLoS One **11**(10), e0163477 (2016). https://doi.org/10.1371/journal.pone.0163477
62. D. Zhang, S. Chen, Z.H. Zhou, Entropy-inspired competitive clustering algorithms. Int. J. Softw. Inform. **1**(1), 67–84 (2007)
63. A. Zimek, E. Schubert, H.P. Kriegel, A survey on unsupervised outlier detection in high-dimensional numerical data. Stat. Anal. Data Min. ASA Data Sci. J. **5**(5), 363–387 (2012). https://doi.org/10.1002/sam.11161

Chapter 4
An Introduction to Deep Clustering

Gopi Chand Nutakki, Behnoush Abdollahi, Wenlong Sun, and Olfa Nasraoui

4.1 Introduction

Driven by the explosive growth in available data and decreasing costs of computation, Deep Learning (DL) has been paving a transformational path in machine learning and artificial intelligence [8, 17], with a dramatic impact on a variety of application domains, ranging from computer vision [15] and speech recognition [13] to natural language processing [5] and recommender systems [25, 27, 30]. DL found much of its fame in problems involving predictive modeling tasks such as classification and recommendation which are considered supervised learning. Deep learning has also been widely used to learn richer and better data representations from big data, without relying too much on human engineered features. Many of these deep representation networks rely on a preliminary unsupervised learning stage, referred to as unsupervised pretraining (e.g., autoencoders, matrix factorization, restricted Boltzmann machines, etc.), which learn better (deep) representations of the data that are known to drastically improve the results of supervised learning networks. Even though it started mostly within the realm of supervised learning, deep learning's success has recently inspired several deep learning-based developments in clustering algorithms which sit squarely within unsupervised learning. Most DL-based clustering approaches have exploited the representational power of DL networks for preprocessing clustering inputs in a way to improve the quality of clustering results and to make clusters easier to extract. However, clustering has not

G. C. Nutakki · B. Abdollahi · W. Sun
Knowledge Discovery & Web Mining Lab, University of Louisville, Louisville, KY, USA
e-mail: g0nuta01@louisville.edu; b0abdo03@louisville.edu; w0sun005@louisville.edu

O. Nasraoui (✉)
Department of Computer Engineering and Computer Science, University of Louisville, Louisville, KY, USA
e-mail: olfa.nasraoui@louisville.edu

O. Nasraoui, C.-E. Ben N'Cir (eds.), *Clustering Methods for Big Data Analytics*, Unsupervised and Semi-Supervised Learning,
https://doi.org/10.1007/978-3-319-97864-2_4

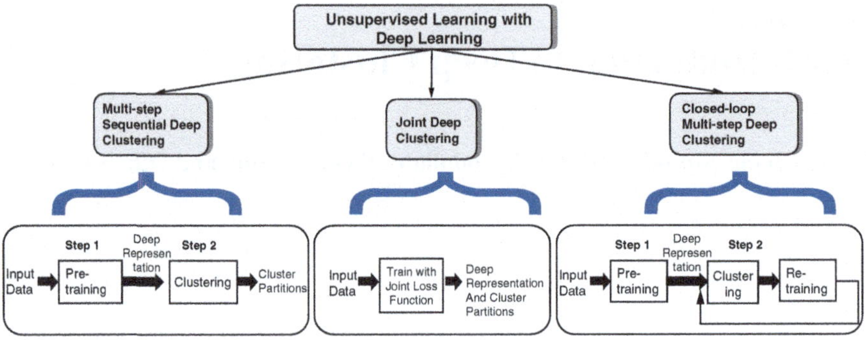

Fig. 4.1 Taxonomy of Deep Clustering presented in this chapter

always been the ultimate goal of these techniques, since some of them aim primarily to obtain richer deep representations, but employ clustering as an essential and enabling step to improve this representation. Yet, both approaches result in deep representations and clustering outputs, hence we will refer to all the approaches presented in this chapter as *Deep Clustering*.

Figure 4.1 helps further understand the taxonomy of Deep Clustering approaches that we present, which itself motivates the structure of this chapter. Deep Clustering approaches differ based on their overall algorithmic structure, network architectures, loss functions, and optimization methods for training (or learning the parameters). In this chapter, we focus on Deep learning for clustering approaches where those approaches either use deep learning for the purpose of grouping (or partitioning) the data and/or creating low rank deep representations or embeddings of the data, which among other possible goals, could play a significant supporting role as a building block of supervised learning. There may be several ways to come up with a taxonomy of deep clustering methods; our approach in this chapter is based on viewing the methods as a *process*. We thus present a simplified taxonomy based on the overall procedural structure or design of deep clustering methods. The simplified taxonomy helps both beginner and advanced readers. Beginners are expected to quickly grasp how almost all approaches are designed based on a small set of common patterns, while more advanced readers are expected to use and extend these patterns in order to design increasingly complex deep clustering pipelines that fit their own machine learning problem-solving aims. In our taxonomy, Deep Clustering approaches can be considered to fall into the following three broad families (see Fig. 4.1):

1. **Sequential multistep Deep Clustering approaches**: These approaches consist of two main steps. The first step learns a richer deep (also known as latent) representation of the input data, while the second step performs clustering on this deep or latent representation.
2. **Joint Deep Clustering approaches**: This family of methods include a step where the representation learning is tightly coupled with the clustering, instead of two separate steps for the representation learning and clustering, respectively.

The tight coupling is generally performed by optimizing a combined or joint loss function that favors good reconstruction while taking into account some form of grouping, clustering, or codebook representation of the data.

3. **Closed-loop multistep Deep Clustering approaches**: This family of methods contains two salient steps, similar to the first family (sequential multistep Deep Clustering); however, the steps alternate in an iterative loop instead of being performed in one feedforward linear fashion.

A common thread between all the above families is that clustering is performed on a different (so-called deep or latent) representation (DR) instead of the original data. The deep representation is typically of lower dimensionality and captures more easily the different hidden groups within the data. It is not surprising therefore that such deep representations are more conducive to better clustering, even using simple algorithms such as K-means although the latter tends to struggle when faced with high-dimensional data. It is well known that using a Deep Representation (DR) or preprocessing of the data using dimensionality reduction methods gives better clustering results. Deep Representation methods that are known to improve the results of K-means include: (1) linear mapping methods such as Principal Component Analysis (PCA) or Nonnegative Matrix Factorization (NMF) [4, 35] and (2) nonlinear approaches such as those used in Spectral Clustering [22, 33] and Deep Neural Network-based DR [3, 10, 12, 28, 32].

Underlying all three families are building blocks which are essential to designing most deep learning methods. These building blocks are:

- Deep representation models.
- Loss functions.

The following sections of the chapter will start by presenting the essential building blocks to Deep Clustering. Then, each family of Deep Clustering approaches will be discussed in order (sequential multistep, joint, and finally closed-loop multistep Deep Clustering).

4.2 Essential Building Blocks for Deep Clustering

Before we introduce deep clustering techniques, we first discuss the essential ingredients for any good clustering. The first important ingredient is that the algorithm must have good representation or features (and/or similarity or affinity measures). The second ingredient is a good cost or loss function that captures what a good representation or clustering is. For this reason, we consider that underlying all three families of Deep Clustering are the following set of building blocks which are essential to designing most deep learning methods:

- Deep representation models: These are unsupervised pretraining models which typically compute new deep or latent features or embeddings that are considered faithful but richer representations (DR) of the data.

- Loss functions: These are the objective or cost functions that are used to train the above deep representation models or to compute clusters from data.

These building blocks are described in the next subsections.

4.2.1 Learning Deep Representations

A high-level representation of features can be learned using deep neural networks (DNNs) that learn mappings from the input data to better (deep) representations (DR) that can in turn aid clustering algorithms to discover better partitions or cluster parameters. The features of these deep representations are generally extracted from one layer of the network, typically the latent hidden layer of an Autoencoder [31] or a Restricted Boltzmann Machine network [11]. The features can also be extracted from the concatenation of several layers [26].

Deep neural networks that can discover better feature mappings are typically trained in a similar way as unsupervised pretraining which is a common preliminary stage even in building supervised deep learning networks. Unsupervised pretraining strategies include denoising autoencoders [31] and Restricted Boltzmann Machines [11] which are known to reduce variance and learn hidden units that compute features of the input data that correspond to major factors of variation in the true input distribution. Hence, unsupervised pretraining can guide optimization towards basins of attraction of minima that allow better generalization from training data and add robustness to a deep architecture [6]. Autoencoders are in fact fundamentally connected to clustering as shown by Baldi [2] who presented a framework to study linear and nonlinear autoencoders, showing that learning in the Boolean autoencoder (which is the most nonlinear autoencoder) is equivalent to a clustering problem that can be solved in polynomial time for a small number of clusters, but becomes NP complete for a large number of clusters.

In addition to autoencoders, there are alternative representation extraction building blocks such as Matrix Factorization [4, 34]. Although linear, Matrix Factorization has proven to be a powerful strategy to learn latent representations that are useful in many applications. MF methods are no longer confined to be linear. In fact, a nonlinear extension of MF, called Generalized Matrix Factorization, was recently proposed and used in deep recommender systems [9]. Just like autoencoders and Restricted Boltzman Machines, they can easily be stacked to form hierarchical representation layers whose output is later used as DR for clustering or other tasks.

4.2.2 Deep Clustering Loss Functions

There are several types of loss functions that can be used in a Deep Clustering framework, of which we describe the four most important types ([1] discussed more loss function building blocks). The first type comes purely from learning a deep representation using deep learning techniques, and independent of any clustering

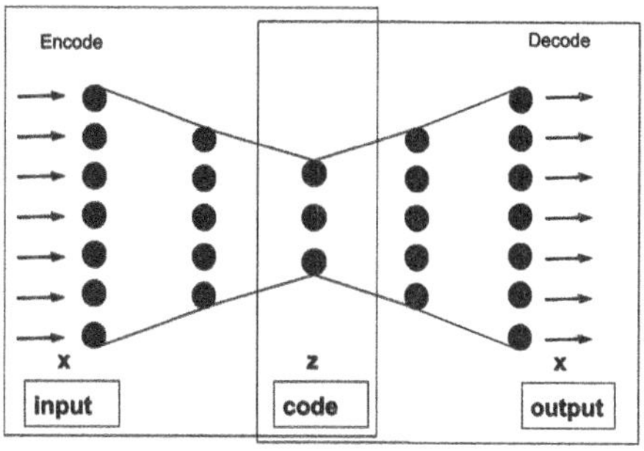

Fig. 4.2 The overall structure of an auto-encoder

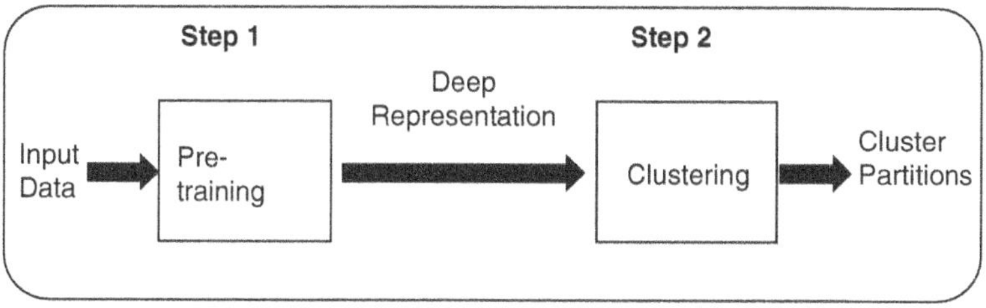

Fig. 4.3 Sequential Multistep Deep Clustering

(e.g., the reconstruction error of an autoencoder or matrix factorization). The second type is the *clustering loss*, which comes from a clustering process (e.g., the K-means' sum of squared errors objective function). The third type is a *joint loss function* that combines the first two types. The fourth type is a loss function meant to obtain crisper or harder cluster partitions in case of soft clustering. Soft or fuzzy clustering allows a data point to belong to multiple clusters with varying degrees of membership.

Autoencoder Reconstruction Loss An autoencoder consists of two parts: an encoder and a decoder [31] (see Fig. 4.3). The encoder maps (or encodes) its input data layer x to a hidden layer representation or a code z in a latent space Z. During training, the decoder tries to learn to make a faithful reconstruction x from the code z, making sure that useful information has not been lost by the encoding process. Once the training is done, the decoder part is no longer used, and only the encoder part (consisting of the input layer and hidden layer) is left for mapping its input to the latent space Z. This procedure allows autoencoders to learn useful representations in case the output's dimensionality is different from that of the inputs or when random noise is injected into the input [32]. They can also be used for dimensionality reduction [12]. The Autoencoder's reconstruction loss captures the distance or error between the input x_i to the autoencoder and the corresponding reconstruction or code $f(x_i)$, for instance, the mean squared error:

$$L = \sum_i \|x_i - f_{autoencoder}(x)\|^2. \tag{4.1}$$

Clustering Loss This type of loss is the objective function that guides a clustering process, such as K-means [20]. Given a set of data samples $\{\boldsymbol{x}_i\}_{i=1,\ldots,N}$ where $\boldsymbol{x}_i \in \mathbb{R}^M$, K-means strives to group the N data samples into K categories or cluster partitions of the data. This is done by minimizing the following loss or cost function:

$$\min_{\boldsymbol{M}\in\mathbb{R}^{M\times K},\{\boldsymbol{s}_i\in\mathbb{R}^K\}} \sum_{i=1}^{N} \|\boldsymbol{x}_i - \boldsymbol{M}\boldsymbol{s}_i\|_2^2 \tag{4.2}$$

$$\text{s.t. } s_{j,i} \in \{0, 1\}, \ \mathbf{1}^T \boldsymbol{s}_i = 1 \ \forall i, j,$$

where $\boldsymbol{s}_i$ is the cluster membership assignment vector of data point i which has only one nonzero element for the cluster nearest to the data point; $s_{j,i}$ is the jth element of $\boldsymbol{s}_i$ and represents the membership assignment in the jth cluster (1 if this cluster is the nearest cluster to the data point and 0 otherwise); and the kth column of $\boldsymbol{M}$, $\boldsymbol{m}_k$, is the centroid of the kth cluster.

Joint Deep Clustering Loss Function A joint loss function is intended to jointly learn both DNN parameters and cluster parameters that aim to produce both better DR and better clusters. Hence, joint loss functions combine both DR and clustering objectives [23, 40]. Joint loss functions typically assume a generative model that generates an M-dimensional data sample by $\boldsymbol{x}_i = \boldsymbol{W}\boldsymbol{h}_i$, where $\boldsymbol{h}_i \in \mathbb{R}^R$ are latent factors and $\boldsymbol{W} \in \mathbb{R}^{M\times R}$ are encoding weights or loading coefficients from the data on those factors, and the latent factors' dimensionality $R \ll M$. They further assume that the data clusters are well-separated in latent domain (i.e., where $\boldsymbol{h}_i$ lives) but distorted by the transformation introduced by $\boldsymbol{W}$. A joint loss function can then be defined as follows [40]:

$$\min_{\boldsymbol{M},\{\boldsymbol{s}_i\},\boldsymbol{W},\boldsymbol{H}} \|\boldsymbol{X} - \boldsymbol{W}\boldsymbol{H}\|_F^2 + \lambda \sum_{i=1}^{N} \|\boldsymbol{h}_i - \boldsymbol{M}\boldsymbol{s}_i\|_2^2$$

$$+ r_1(\boldsymbol{H}) + r_2(\boldsymbol{W}) \tag{4.3}$$

$$\text{s.t. } s_{j,i} \in \{0, 1\}, \ \mathbf{1}^T \boldsymbol{s}_i = 1 \ \forall i, j,$$

where $\boldsymbol{X} = [\boldsymbol{x}_1, \ldots, \boldsymbol{x}_N]$, $\boldsymbol{H} = [\boldsymbol{h}_1, \ldots, \boldsymbol{h}_N]$, and $\lambda \geq 0$ is a parameter for balancing data fidelity or reconstruction (first term) and the latent cluster structure (second term). In (4.3), the first term performs DR and the second term performs latent clustering. The terms $r_1(\cdot)$ and $r_2(\cdot)$ are regularization terms that try to favor nonnegativity or sparsity and are used to prevent trivial solutions, such as zero latent factor values [37].

Cluster Assignment Hardening Loss Assuming soft assignments of data points to clusters, obtained using fuzzy clustering techniques, probabilistic clustering methods, or normalized similarities between points and centroids. The *cluster assignment hardening* loss tries to enforce making soft assignment probabilities stricter or harder by encouraging cluster assignment probability distribution Q to approach an auxiliary (target) distribution P which guarantees this constraint. The following auxiliary distribution [34] improves cluster purity and stresses data points that are assigned with high confidence:

$$p_{ij} = \frac{q_{ij}^2 / \Sigma_i q_{ij}}{\Sigma_{j'} (q_{ij'}^2 / \Sigma_i q_{ij'})}. \tag{4.4}$$

because it forces assignments to have stricter probabilities (closer to 0 and 1). This happens due to squaring the original distribution and then normalizing. Cluster hardening can finally be accomplished by minimizing the distance between the original cluster assignment probability distribution and the auxiliary or target crisp distribution, which can be done by minimizing (via neural network training) their Kullback–Leibler divergence [16], given by:

$$L = \mathrm{KL}(P \| Q) = \sum_i \sum_j p_{ij} \log \frac{p_{ij}}{q_{ij}}. \tag{4.5}$$

4.3 Sequential Multistep Deep Clustering

Sequential multistep Deep Clustering approaches consist of two major steps, as shown in Fig. 4.3. The first step learns a richer deep (also known as latent) representation (DR) of the input data, while the second step performs clustering on this deep or latent representation. In the following, we describe a few representative algorithms that fall into this family of Deep Clustering methods.

4.3.1 Fast Spectral Clustering

Fast Spectral Clustering [38] is a classical multistep approach where an autoencoder is trained to produce a latent or deep representation (DR), which is later used as input to clustering with the K-means algorithm. In contrast to multistep methods that train the autoencoder directly on the input data, Fast Spectral Clustering first extracts an embedding S from the Laplacian of a precomputed similarity or kernel matrix W and then trains the autoencoder to encode this embedding (see Algorithm 1). The similarities are computed using a Gaussian kernel between every input data point and p landmark points. The p landmarks can be obtained in different ways. They

Algorithm 1 Fast Spectral Clustering (FSC)

Input: Input data $\{\mathbf{x}_1, \mathbf{x}_2, ..., \mathbf{x}_n\}$;
Output: k clusters of the data set 1: Select p landmarks 2: Compute similarity matrix W between data points and landmarks 3: Compute the degree matrix: $D = diag(W^\top \mathbf{w}^s)$ 4: Compute S, the input to the autoencoder: $\mathbf{s}_i = d_{ii}^{-1/2} \mathbf{w}_i$ 5: Train an autoencoder using S as its input 6: Run k-means on the latent space DR of the trained autoencoder

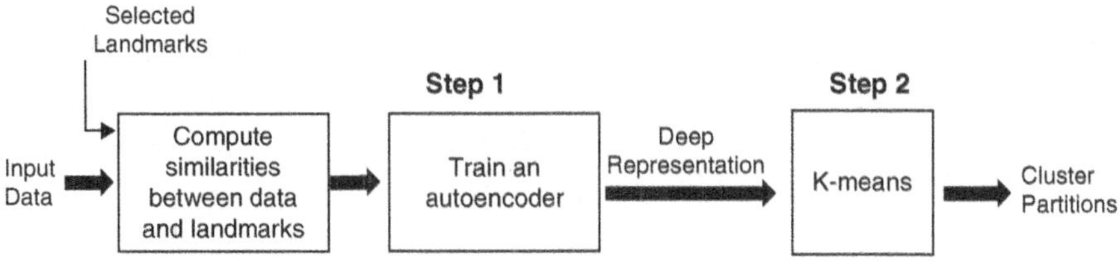

Fig. 4.4 Fast Spectral Clustering

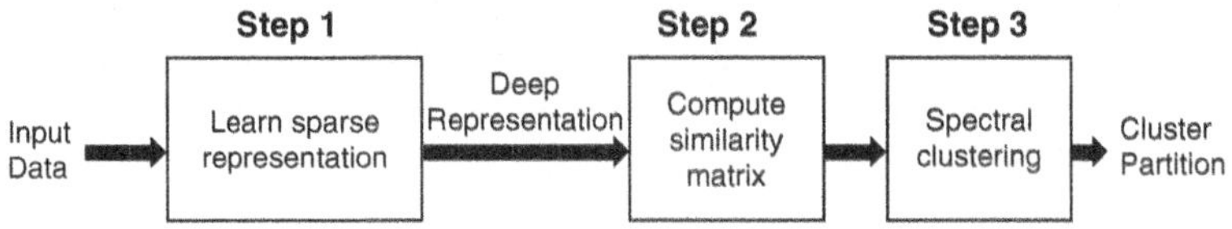

Fig. 4.5 Deep Sparse Subspace Clustering

can be randomly sampled from the original data set or selected from the centroids of p clusters by running k-means or using column subset selection methods [7]. The Laplacian matrix is computed using:

$$Ł_{norm} = D^{-1/2} M D^{-1/2} = D^{-1/2} W^{\top} W D^{-1/2} = S \cdot S^T \tag{4.6}$$

which yields the embedding $S = W D^{-1/2}$ that will be used as the input to the autoencoder (Fig. 4.4).

4.3.2 Deep Sparse Subspace Clustering (SSC)

Given a data set $X = [x_1, x_2, \ldots, x_n] \in R^{d \times n}$, SSC [24] seeks to linearly reconstruct the i-th sample $\mathbf{x}_i$ using a small number of other samples, hence producing representation coefficients that are sparse. The optimization consists of minimizing the reconstruction loss function as follows:

$$\min_{\mathbf{c}_i} \frac{1}{2} \|\mathbf{x}_i - \mathbf{X}\mathbf{c}_i\|_F^2 + \gamma \|\mathbf{c}_i\|_1 \qquad \text{s.t. } c_{ii} = 0 \tag{4.7}$$

where $\| \cdot \|_1$ denotes the ℓ_1-norm and c_{ii} denotes the i-th element in $\mathbf{c}_i$ which encourages $\mathbf{c}_i$ to be sparse, and the constraint $c_{ii} = 0$ avoids trivial solutions. After solving for the sparse representation $\mathbf{C}$, an affinity matrix $\mathbf{A}$ is calculated using $\mathbf{A} = |\mathbf{C}| + |\mathbf{C}|^T$. This matrix is finally used as input to spectral clustering to obtain clustering results (see Fig. 4.5).

4.3.3 Deep Subspace Clustering (DSC)

In most existing subspace clustering methods including SSC, each data input is encoded as a linear combination of the whole data set. However, linear combinations may not be sufficient for representing high-dimensional data which usually lie on nonlinear manifolds. Deep Subspace Clustering (DSC) [24] represents each input sample as a combination of others, similar to SCC. However, instead of a linear combination, DSC learns a mapping using explicit hierarchical transformations in a neural network and simultaneously learns the reconstruction coefficients **C**. The neural network consists of $M + 1$ stacked layers with M nonlinear transformations, which takes a given data input x as the input to the first layer. The input to the first layer of the neural network is denoted as $h^{(0)} = x \in R^d$, while $h^{(m)}$ denotes the output of the subsequent (m-th) layers (where $m = 1, 2, \ldots, M$ indexes the layers), $W^{(m)} \in R^{d^{(m)} \times d^{(m-1)}}$ and $b^{(m)} \in R^{d^{(m)}}$ denote the weights and bias associated with the m-th layer, respectively, and $d^{(m)}$ is the dimension of the output of the m-th layer. If $H^{(M)}$ denotes the collection of the corresponding outputs given by the neural network,

$$\mathbf{H}^{(M)} = [\mathbf{h}_1^{(M)}, \mathbf{h}_2^{(M)}, \ldots, \mathbf{h}_n^{(M)}], \tag{4.8}$$

the optimization problem of DSSC can then be expressed as:

$$\begin{aligned} \min_{\mathbf{W}^{(m)}, \mathbf{b}^{(m)}, \mathbf{C}} \; & \frac{1}{2} \|\mathbf{H}^{(M)} - \mathbf{H}^{(M)}\mathbf{C}\|_F^2 + \gamma \|\mathbf{C}\|_1 \\ & + \frac{\lambda}{4} \sum_{i=1}^{n} \|(\mathbf{h}_i^{(M)})^T \mathbf{h}_i^{(M)} - 1\|_2^2 \\ & \text{s.t.} \, diag(\mathbf{C}) = 0, \end{aligned} \tag{4.9}$$

where λ is a positive trade-off parameter, the first term is designed to minimize the discrepancy between $\mathbf{H}^{(M)}$ and its reconstructed representation, the second term regularizes **C** for some desired properties, and the third term is designed to remove an arbitrary scaling factor in the latent space, without which the neural network may collapse in the trivial solutions $\mathbf{H}^{(M)} = \mathbf{0}$.

DSSC simultaneously learns M nonlinear mapping functions $\{\mathbf{W}^{(m)}, \mathbf{b}^{(m)}\}_{m=1}^M$ and n sparse codes $\{\mathbf{c}_i\}_{i=1}^n$ by alternatively updating one variable while fixing all others, until convergence. Stochastic sub-gradient descent (SGD) is used to update the parameters $\{\mathbf{W}^{(m)}, \mathbf{b}^{(m)}\}_{m=1}^M$. An ℓ_2-norm regularizer can also be added to avoid overfitting [15, 21]. After obtaining **C**, a similarity graph is constructed using $\mathbf{A} = |\mathbf{C}| + |\mathbf{C}|^T$ and this graph is used as input to clustering.

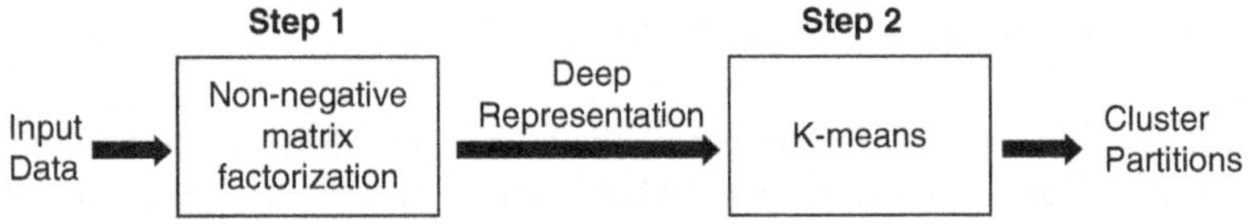

Fig. 4.6 Nonnegative matrix factorization (NMF) + K-means

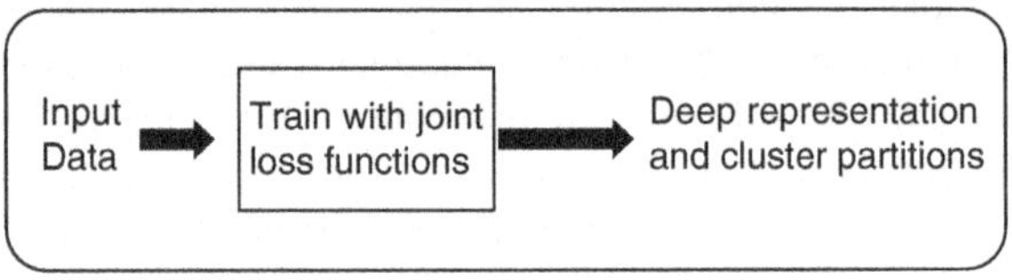

Fig. 4.7 Joint deep clustering

4.3.4 Nonnegative Matrix Factorization (NMF) + K-Means

This approach first applies NMF [18] to learn a latent or deep representation, and then applies K-means to this reduced-dimension representation [39] (see Fig. 4.6). It is a common baseline for evaluating more complex architectures.

4.4 Joint Deep Clustering

Joint Deep Clustering is a family of methods that include a step, where the representation learning is tightly coupled with the clustering, instead of two separate steps for the representation learning and clustering, respectively. The tight coupling is generally performed by optimizing a combined or joint loss function that favors good reconstruction while taking into account some form of grouping, clustering, or codebook representation of the data. In the following, we describe a few representative algorithms that fall into this family of Deep Clustering methods (Fig. 4.7).

4.4.1 Task-Specific and Graph-Regularized Network (TAGnet)

The goal of TAGnet [41] is to learn a discriminative embedding that is optimal for clustering. Different from generic deep architectures, TAGnet is designed in a way to take advantage of the successful sparse code-based clustering pipelines. The TAGnet approach includes a feed-forward architecture, termed *Task-specific And Graph-regularized Network* (TAGnet), to learn discriminative features, and a joint clustering-oriented loss function (see Fig. 4.8). Its aim is to learn features that are optimized under clustering criteria, while encoding graph constraints to regularize the target solution. The solution to training the network parameters in TAGnet is

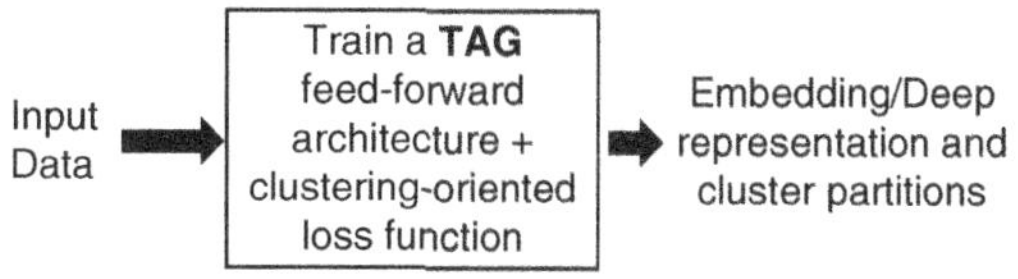

Fig. 4.8 The TAGnet approach

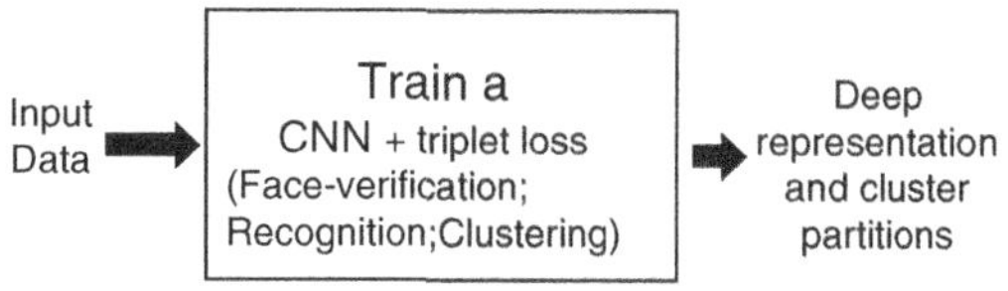

Fig. 4.9 The FaceNet approach

derived from a theorem that shows that the optimal sparse code can be obtained as the fixed point of an iterative application of a shrinkage operation on the degree matrix and the Laplacian derived from the graph's affinity matrix.

4.4.2 FaceNet

Designed for face recognition tasks, FaceNet [28] is a unified system for face verification, recognition, and clustering (see Fig. 4.9). The approach aims to learn a Euclidean embedding for images using a deep convolutional network. The network is trained to produce good embeddings such that the squared L2 distances in the embedding space directly correspond to the notion of face similarity; thus, faces of the same person should have small distances while faces of different people should be separated by large distances. Training is achieved by minimizing a triplet loss function that is used to simultaneously achieve face verification, recognition, and clustering. An embedding $f(x)$ is extracted from an image x into a feature space $\mathbb{R}^d$, where the squared distance between all faces of the same identity is small across all imaging conditions, whereas the squared distance between a pair of face images from different identities is large.

4.4.3 Deep Clustering Network (DCN)

Deep Clustering Network (DCN) [39] is a multistep approach (see Fig. 4.14) that starts by pretraining an autoencoder based on reconstruction loss minimization, then feeds the deep representation output to K-means for clustering. After clustering, the autoencoder is retrained by minimizing a joint loss function combining reconstruction loss and the K-means clustering loss. DCN then alternates between the autoencoder network training and cluster updates. DCN's results on the MNIST data set outperformed the results of the similar DEC approach, presented earlier (Fig. 4.10).

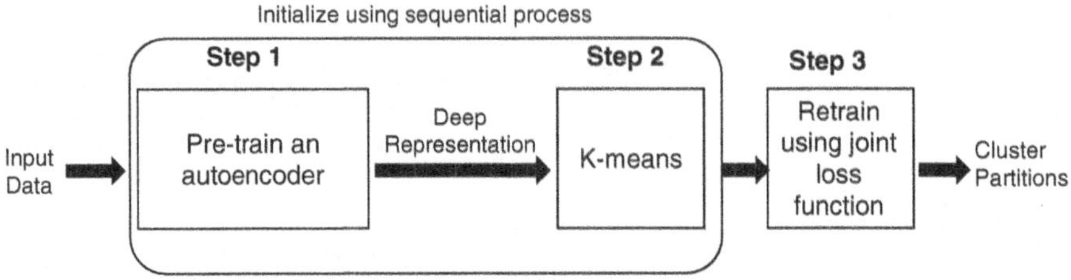

Fig. 4.10 Overview of a Deep Clustering Network (DCN) model

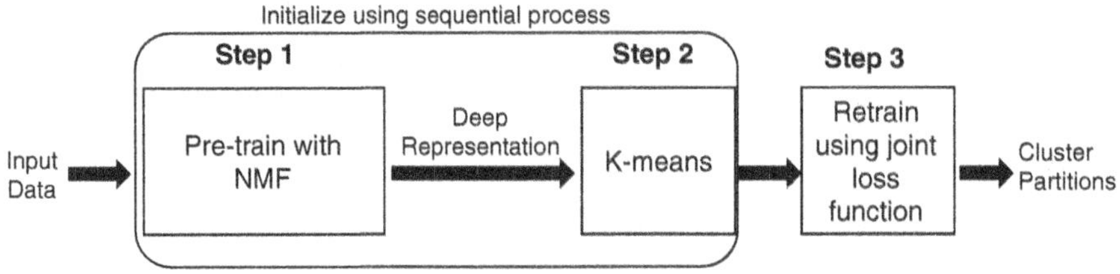

Fig. 4.11 Joint NMF + K-means (JNKM) model

4.4.4 Joint NMF and K-Means (JNKM)

Similar to DCN, JNKM [39] performs joint deep representation and K-means clustering in several steps, except that it uses NMF instead of an autoencoder to obtain the DR. This method has the same structure as DCN (see Fig. 4.11), replacing the autoencoder training with NMF, and was used as a baseline by [39]. Recall that the main difference between NMF and autoencoder networks is the nonlinearity of the latter's mapping.

4.5 Closed-Loop Multistep Deep Clustering

Closed-loop multistep Deep Clustering approaches are a family of methods that consist of two salient steps, similar to the first family (sequential multistep Deep Clustering); however, these steps alternate in a loop instead of being performed in one feedforward linear fashion. In the following, we describe a few representative algorithms that fall into this family of Deep Clustering methods (Fig. 4.12).

Deep Embedded Clustering (DEC) uses autoencoders as network architecture and initialization method, and uses K-means for clustering [34] (see Fig. 4.13). DEC first pretrains an autoencoder using a standard input reconstruction loss function. Then, the DR resulting from this autoencoder is fed to K-means to obtain clusters. Next, the autoencoder network is fine-tuned using the cluster assignment hardening loss (see the Building Blocks section above) and the clustering centers

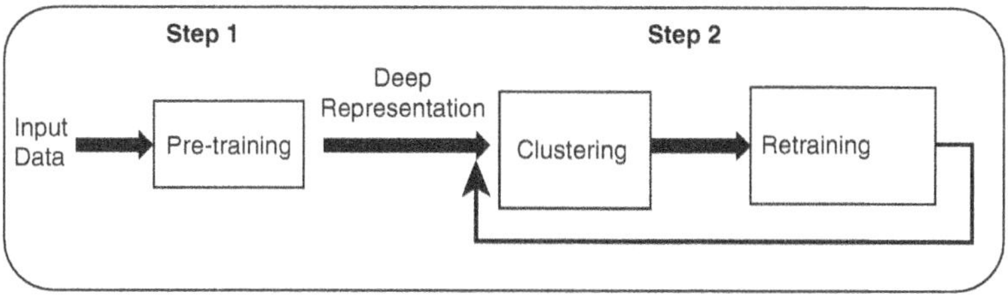

Fig. 4.12 Closed-loop multistep deep clustering

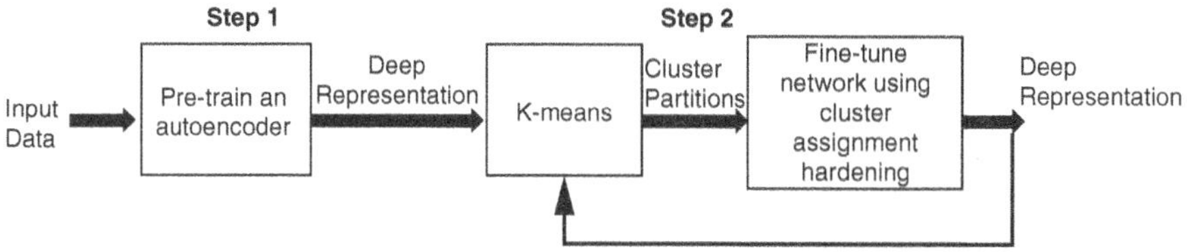

Fig. 4.13 Deep Embedded Clustering (DEC)

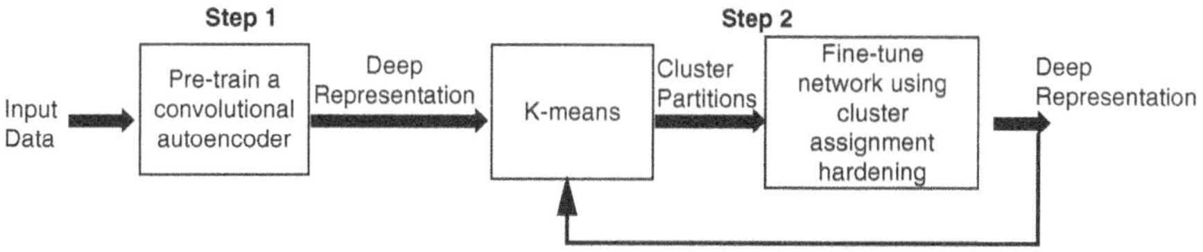

Fig. 4.14 Discriminatively Boosted Clustering (DBC)

are updated again. This process is then repeated. The clusters are iteratively refined by learning from their high confidence assignments with the help of the auxiliary target distribution.

Discriminatively Boosted Clustering (DBC) [19] shares many similarities with DEC except for using convolutional autoencoders. It also uses K-means for clustering and the same training method: pretraining with autoencoder reconstruction loss, clustering the autoencoder's output DR, then fine-tuning the autoencoder using a cluster assignment hardening loss (see Fig. 4.14). The main advantage of DBC over DEC is its superior performance on image data, as expected, due to using convolutional layers.

Clustering CNN (CCNN) uses a Convolutional Neural Network [29] to achieve joint clustering and deep representation learning. The features are extracted from one of the internal layers of the CCNN, while the cluster labels are considered to be the output predictions coming from the CCNN's softmax layer. The initial cluster centers that are used to compute the joint loss function in the pretraining phase are based on selecting k random images from the data set. After clustering the features coming from the DR using K-means, the assigned cluster labels are compared with the labels predicted by the CCN's softmax layer to compute a

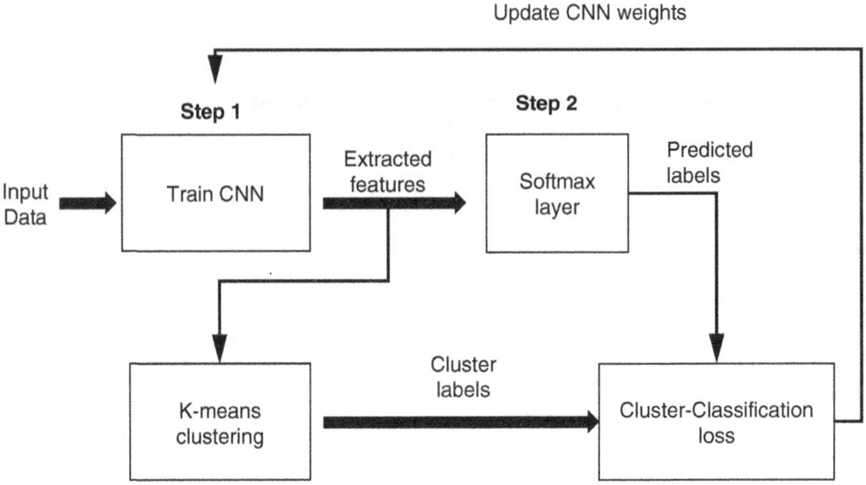

Fig. 4.15 Clustering Convolutional Neural Network (CCNN)

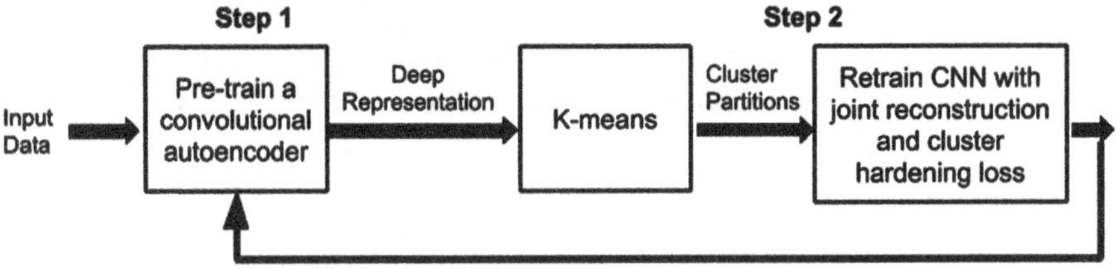

Fig. 4.16 Aljalbout's closed loop approach using a joint loss function

clustering-classification loss. This loss function is then used to retrain the CCNN parameters. The process is repeated with the extracted DR features again used to update the cluster centroids, essentially entering a repetitive loop of CCN training and clustering (see Fig. 4.15).

Aljalbout et al. [1] proposed a closed loop strategy that also iterates autoencoder training and K-Means clustering in a loop. Like CCNN, they used a convolutional architecture for extracting the DR in the first step to handle image data. However, the main difference is that they retrain the network weights using a joint loss function combining both reconstruction and cluster hardening loss (see Fig. 4.16). Their method was shown to outperform other strategies on the MNIST and COIL20 benchmark image data sets based on the external cluster evaluation metrics, Normalized Mutual Information and cluster accuracy [1].

4.6 Conclusions

In this chapter, we presented a simplified taxonomy with several representative examples of Deep Clustering methods which are built on algorithms that result in both deep or latent representations (DR) and (either as an explicit aim or as a byproduct) clustering outputs, such as a partition of the input data and cluster centroids. Our simplified taxonomy is based on the overall procedural structure or design of most Deep Clustering methods, thus dividing Deep Clustering approaches into three families: sequential multistep, joint, and closed-loop multistep Deep Clustering methods. In addition to describing several algorithms that fall under the three main families above, we discussed the most important building blocks that are essential to Deep Clustering methods.

Together with the Deep Clustering building blocks, our simplified taxonomy should help beginning readers get a quick grasp of how most approaches are designed, while guiding more advanced readers, and following the tradition of the DL community, to stack and combine the modular patterns that we have presented in order to design even more sophisticated deep clustering pipelines for their specific problem-solving needs.

While we have not detailed the implementation details or computational costs of Deep Clustering, scaling Deep Clustering follows along the steps of scaling DL in general. This means that strategies such as SGD play an essential role in fast optimization in the presence of big data sets, while GPU clusters play a critical role in accelerating computation due to the highly distributed nature of most DL architectures and weight updates. As with conventional clustering, graph-based, agglomerative, and spectral methods are notoriously costly in both computation time and memory requirements because of handling pairwise similarity matrices and Laplacians. However, this can be addressed by using a small number of selected landmarks instead of the entire data set when computing pairwise metrics, such as was done in Fast Spectral Clustering [38].

Deep Clustering has been applied to several big data domains, ranging from image to text and time series (earthquake) data, and we expect that in the future, it will have an impact on even more diverse domains and applications, in the same tradition of older clustering algorithms. Deep Clustering is hence expected to continue the tradition of clustering algorithms and to expand their ability to elucidate the hidden structure in big data and thus contribute to better understanding, retrieval, visualization, and organization of big data, in addition to being an important component of complex decision-making systems [14, 36].

References

1. E. Aljalbout, V. Golkov, Y. Siddiqui, D. Cremers, Clustering with deep learning: taxonomy and new methods (2018). Arxiv preprint arXiv:1801.07648
2. P. Baldi, Autoencoders, unsupervised learning, and deep architectures, in *Proceedings of ICML Workshop on Unsupervised and Transfer Learning* (2012)

3. J. Bruna, S. Mallat, Invariant scattering convolution networks. IEEE Trans. Pattern Anal. Mach. Intell. **35**(8), 1872–1886 (2013)
4. D. Cai, X. He, J. Han, Locally consistent concept factorization for document clustering. IEEE Trans. Knowl. Data Eng. **23**(6), 902–913 (2011)
5. R. Collobert, J. Weston, A unified architecture for natural language processing: deep neural networks with multitask learning, in *Proceedings of the 25th International Conference on Machine Learning* (ACM, New York, 2008), pp. 160–167
6. D. Erhan, Y. Bengio, A. Courville, P.-A. Manzagol, P. Vincent, S. Bengio, Why does unsupervised pre-training help deep learning? J. Mach. Learn. Res. **11**, 625–660 (2010)
7. A.K. Farahat, A. Elgohary, A. Ghodsi, M.S. Kamel, Greedy column subset selection for large-scale data sets. Knowl. Inf. Syst. **45**(1), 1–34 (2015)
8. I. Goodfellow, Y. Bengio, A. Courville, Y. Bengio, *Deep Learning*, vol. 1 (MIT Press, Cambridge, 2016)
9. X. He, L. Liao, H. Zhang, L. Nie, X. Hu, T.-S. Chua, Neural collaborative filtering, in *Proceedings of the 26th International Conference on World Wide Web*. International World Wide Web Conferences Steering Committee (2017), pp. 173–182
10. J.R. Hershey, Z. Chen, J. Le Roux, S. Watanabe, Deep clustering: discriminative embeddings for segmentation and separation, in *2016 IEEE International Conference on Acoustics, Speech and Signal Processing (ICASSP)* (IEEE, Piscataway, 2016), pp. 31–35
11. G.E. Hinton, A practical guide to training restricted Boltzmann machines, in *Neural Networks: Tricks of the Trade* (Springer, Berlin, 2012), pp. 599–619
12. G.E. Hinton, R.R. Salakhutdinov, Reducing the dimensionality of data with neural networks. Science **313**(5786), 504–507 (2006)
13. G. Hinton, L. Deng, D. Yu, G.E. Dahl, A.-R. Mohamed, N. Jaitly, A. Senior, V. Vanhoucke, P. Nguyen, T.N. Sainath et al., Deep neural networks for acoustic modeling in speech recognition: the shared views of four research groups. IEEE Signal Process. Mag. **29**(6), 82–97 (2012)
14. A.K. Jain, Data clustering: 50 years beyond k-means. Pattern Recogn. Lett. **31**(8), 651–666 (2010)
15. A. Krizhevsky, I. Sutskever, G.E. Hinton, ImageNet classification with deep convolutional neural networks, in *Advances in Neural Information Processing Systems* (2012), pp. 1097–1105
16. S. Kullback, R.A. Leibler, On information and sufficiency. Ann. Math. Stat. **22**(1), 79–86 (1951)
17. Y. LeCun, Y. Bengio, G. Hinton, Deep learning. Nature **521**(7553), 436 (2015)
18. D.D. Lee, H.S. Seung, Learning the parts of objects by non-negative matrix factorization. Nature **401**(6755), 788 (1999)
19. F. Li, H. Qiao, B. Zhang, X. Xi, Discriminatively boosted image clustering with fully convolutional auto-encoders (2017). Arxiv preprint arXiv:1703.07980
20. S. Lloyd, Least squares quantization in PCM. IEEE Trans. Inf. Theory **28**(2), 129–137 (1982)
21. G. Montavon, K.-R. Müller, Better representations: invariant, disentangled and reusable, in *Neural Networks: Tricks of the Trade* (Springer, Berlin, 2012), pp. 559–560
22. A.Y. Ng, M.I. Jordan, Y. Weiss, On spectral clustering: analysis and an algorithm, in *Advances in Neural Information Processing Systems* (2002), pp. 849–856
23. V.M. Patel, H. Van Nguyen, R. Vidal, Latent space sparse subspace clustering, in *2013 IEEE International Conference on Computer Vision (ICCV)* (IEEE, Piscataway, 2013), pp. 225–232
24. X. Peng, J. Feng, S. Xiao, J. Lu, Z. Yi, S. Yan, Deep sparse subspace clustering (2017). ArXiv preprint arXiv:1709.08374
25. M. Quadrana, A. Karatzoglou, B. Hidasi, P. Cremonesi, Personalizing session-based recommendations with hierarchical recurrent neural networks, in *Proceedings of the Eleventh ACM Conference on Recommender Systems* (ACM, New York, 2017), pp. 130–137
26. S. Saito, L. Wei, L. Hu, K. Nagano, H. Li, Photorealistic facial texture inference using deep neural networks, in *IEEE Conference on Computer Vision and Pattern Recognition, CVPR*, vol. 3 (2017)

27. R. Salakhutdinov, A. Mnih, G. Hinton, Restricted Boltzmann machines for collaborative filtering, in *Proceedings of the 24th International Conference on Machine Learning* (ACM, New York, 2007), pp. 791–798
28. F. Schroff, D. Kalenichenko, J. Philbin, FaceNet: a unified embedding for face recognition and clustering, in *Proceedings of the IEEE Conference on Computer Vision and Pattern Recognition* (2015)
29. C.-J. Tsai, Y.-W. Tsai, S.-L. Hsu, Y.-C. Wu, Synthetic training of deep CNN for 3d hand gesture identification, in *2017 International Conference on Control, Artificial Intelligence, Robotics & Optimization (ICCAIRO)* (IEEE, Piscataway, 2017), pp. 165–170
30. A. Van den Oord, S. Dieleman, B. Schrauwen, Deep content-based music recommendation, in *Advances in Neural Information Processing Systems* (2013), pp. 2643–2651
31. P. Vincent, H. Larochelle, Y. Bengio, P.-A. Manzagol, Extracting and composing robust features with denoising autoencoders, in *Proceedings of the 25th International Conference on Machine Learning* (ACM, New York, 2008), pp. 1096–1103
32. P. Vincent, H. Larochelle, I. Lajoie, Y. Bengio, P.-A. Manzagol, Stacked denoising autoencoders: learning useful representations in a deep network with a local denoising criterion. J. Mach. Learn. Res. **11**, 3371–3408 (2010)
33. U. Von Luxburg, A tutorial on spectral clustering. Stat. Comput. **17**(4), 395–416 (2007)
34. J. Xie, R. Girshick, A. Farhadi, Unsupervised deep embedding for clustering analysis, in *International Conference on Machine Learning* (2016), pp. 478–487
35. R. Xu, Measuring explained variation in linear mixed effects models. Stat. Med. **22**(22), 3527–3541 (2003)
36. D. Xu, Y. Tian, A comprehensive survey of clustering algorithms. Ann. Data Sci. **2**(2), 165–193 (2015)
37. Y. Yang, M.C. Gupta, K.L. Dudley, Towards cost-efficient EMI shielding materials using carbon nanostructure-based nanocomposites. Nanotechnology **18**(34), 345701 (2007)
38. A.Y. Yang, S.S. Sastry, A. Ganesh, Y. Ma, Fast ℓ 1-minimization algorithms and an application in robust face recognition: a review, in *2010 17th IEEE International Conference on Image Processing (ICIP)* (IEEE, Piscataway, 2010), pp. 1849–1852
39. B. Yang, X. Fu, N.D. Sidiropoulos, M. Hong, Towards k-means-friendly spaces: simultaneous deep learning and clustering (2016). Arxiv preprint arXiv:1610.04794
40. J. Yang, D. Parikh, D. Batra, Joint unsupervised learning of deep representations and image clusters, in *Proceedings of the IEEE Conference on Computer Vision and Pattern Recognition* (2016), pp. 5147–5156
41. M. Zheng, J. Bu, C. Chen, C. Wang, L. Zhang, G. Qiu, D. Cai, Graph regularized sparse coding for image representation. IEEE Trans. Image Process. **20**(5), 1327–1336 (2011)

Chapter 5
Spark-Based Design of Clustering Using Particle Swarm Optimization

Mariem Moslah, Mohamed Aymen Ben HajKacem, and Nadia Essoussi

5.1 Introduction

Large volume of data are being collected from different sources and there is a high demand for methods that can efficiently analyze such data referred to as Big data analysis. Big data usually refers to three main dimensions, also called the three Vs [11], which are, respectively, *Volume*, *Variety*, and *Velocity*. Volume refers to the large amount of data, Variety refers to the number of types of data, and Velocity refers to the speed of data processing. Hence, exploring and organizing large-scale data using machine learning techniques becomes an important challenge in Big data analysis.

Clustering is an important technique in machine learning which has been used to look for hidden models, relations, or to summarize data. Technically, clustering aims to organize data into a predetermined number of groups where objects within the same group share some common characteristics. Examples of clustering methods categories are hierarchical methods, density-based methods, grid-based methods, model-based methods, and partitional methods [18]. K-means [14] as one of the partitional clustering methods, it remains the most efficient because of its simplicity and its low computational complexity. However, it is sensitive to the selection of initial cluster centers, as it may converge to suboptimal solutions if the initial cluster centers are not properly chosen [5].

To overcome this weakness, several optimization algorithms were introduced to perform data clustering. Genetic algorithm (GA) which is based on a mutation operator for clustering analysis was proposed [12]. Another approach based on simulated annealing (SA) for data clustering was proposed [3] and more recently

M. Moslah · M. A. B. HajKacem (✉) · N. Essoussi
LARODEC, Institut Supérieur de Gestion de Tunis, Université de Tunis, Le Bardo, Tunisia
e-mail: nadia.essoussi@isg.rnu.tn

O. Nasraoui, C.-E. Ben N'Cir (eds.), *Clustering Methods for Big Data Analytics*, Unsupervised and Semi-Supervised Learning,
https://doi.org/10.1007/978-3-319-97864-2_5

the particle swarm optimization (PSO) was proposed for data clustering [17]. Among these algorithms, particle swarm optimization (PSO), as one of the swarm intelligence algorithms, has gain a great popularity in the last two decades and seemed to be potentially full and fertile research area [15]. In addition, PSO algorithm does not require high computational capacities or a lot of parameters to adjust, compared to genetic algorithms [8]. Although the efficiency of PSO algorithm for data clustering, it does not scale with the increasing volume of data. This is explained by the high computation time to build grouping from large amount of data.

To deal with large-scale data, Aljarah and Ludwig [2] proposed fitting PSO clustering into MapReduce model. However, this method has some considerable shortcomings. The first shortcoming is the result of inherent conflict between MapReduce and PSO clustering. PSO is an iterative algorithm and it requires to perform many iterations for producing optimal results. In contrast, MapReduce has a significant problem with iterative algorithms. At each iteration, the whole data set must be read and written to disks and this results a high input/output (I/O) operations. This significantly degrades the performance of MapReduce-based method. The second shortcoming is inherited from the PSO clustering algorithm. PSO suffers from a low convergence speed when it approaches the global optimum region. According to [1], particles tend to move slowly when they reach the convergence region.

To deal with these issues, we propose in this chapter a new Spark-based **PSO** clustering method, referred to as S-PSO. First, we propose to exploit the benefits provided by Spark, by using in-memory operations to reduce the efficiency of existing MapReduce solutions. Second, we propose a modified version of PSO which executes k-means algorithm when it approaches the global optimum region to accelerate the convergence. The rest of this chapter is organized as follows: Sect. 5.2 presents a background about the basic concepts related to the particle swarm optimization algorithm, MapReduce model, and Spark framework. In Sect. 5.3, existing works related to data clustering using PSO and Big data clustering methods are presented. Section 5.4 presents our proposed method S-PSO. Section 5.5 offers a theoretical analysis of the proposed method. Section 5.6 presents experiments that we have performed to evaluate the efficiency of the proposed method. Section 5.7 presents conclusion and future works.

5.2 Background

This section first presents the particle swarm optimization, followed by the Big data technologies used in this work.

5.2.1 Particle Swarm Optimization

Particle swarm optimization (PSO) was first introduced in 1995 by Kennedy the social psychologist and Eberhart the electrical engineer. At first the algorithm was intended to simulate the social behavior of birds when searching for food. When a bird recognizes a food area, it broadcasts the information to all the swarm. Hence, all the birds follow him and this way they raise the probability of finding the food since it is a collaborative work. Then the behavior of birds within swarms was turned into an intelligent algorithm capable of solving several optimization problems [15].

PSO is a population-based optimization algorithm. It is composed of a swarm of particles where each particle is considered as a potential solution to the optimization problem. Each particle i is characterized at the time t, by the current position $x_i(t)$ in the search space, the velocity $v_i(t)$, the personal best position $pbestP_i(t)$, and fitness value $pbestF_i(t)$. The personal best position represents the best position that the particle has ever achieved throughout its movement, which is defined as follows:

$$pbestP_i(t+1) = \begin{cases} pbestP_i(t) \ if \ f(pbestP_i(t)) <= f(x_i(t+1)) \\ x_i(t+1) \ if \ f(pbestP_i(t)) > f(x_i(t+1)) \end{cases} \tag{5.1}$$

The swarm's best position $gbestP(t)$ corresponds to the best position that the whole swarm has ever seen, which is calculated as follows:

$$gbestP(t+1) = \min\ (f(y), f(gbestP(t))) \tag{5.2}$$

where $y \in \{pbestP_0(t), ..., pbestP_S(t)\}$. Particle's positions and velocities are calculated using the following equations:

$$x_i(t+1) \leftarrow x_i(t) + v_i(t) \tag{5.3}$$

$$v_i(t+1) \leftarrow wv_i(t) + c_1r_1(pbestP_i(t) - x_i(t)) + c_2r_2(gbestP(t) - x_i(t)) \tag{5.4}$$

where w is referred to the inertia weight, $x_i(t)$ is the position of the particle i at the time t, $v_i(t)$ is the velocity of the particle i at the time t, c_1 and c_2 are two acceleration coefficients, and r_1 and r_2 are two random values in the range [0,1]. The main algorithm of PSO is described in Algorithm 1 [10].

Algorithm 1 The main algorithm of PSO

```
1: Input: Z: input data set
2: Output: Particles information
3: Initialize the swarm of particles from Z.
4: while Convergences not reached do
5:    Compute the fitness of particles according to the fitness function to be optimized.
6:    Update each particle's personal best position and fitness value using Equation 5.1.
7:    Update the global best position using Equation 5.2.
8:    Update particle's positions and velocities using Equation 5.3 and 5.4 respectively.
9: end while
```

It is important to note that the convergence is reached when *gbest* does not have significant changes anymore [15].

5.2.2 MapReduce Model

MapReduce is a very well-known programming framework built to ensure the parallel computation and processing of large volume of data. It adopts the method of divide and conquer in which a problem is divided into smaller and less complex sub-problems. Then simultaneously, all sub-problems are separately executed. Once finished, results are merged to provide a final solution to the very big and complex problem [6]. The principal components of the MapReduce model are the map and reduce functions. The map function takes as input key/value pairs (k, v), performs the assigned work, and generates intermediate key/value pairs $[(k', v')]$. An intermediate step known as shuffling step is required to organize each intermediate key with its corresponding values [7].

The reduce function aims to merge all the values corresponding to each intermediate key $(k',[v'])$ to form the final result and output final key/value pairs as $[(k' ,v'')]$[7].

Figure 5.1 outlines the flowchart of MapReduce paradigm. The enormous data set is divided into several chunks, small enough to be fitted into a single machine, each chunk is then assigned to a map function to be processed in parallel. Inputs and outputs are stored in the distributed file system and are accessible by each node in the cluster. Apache Hadoop is the most popular implementation of MapReduce for Big data processing and storage on commodity hardware. The use of this framework has

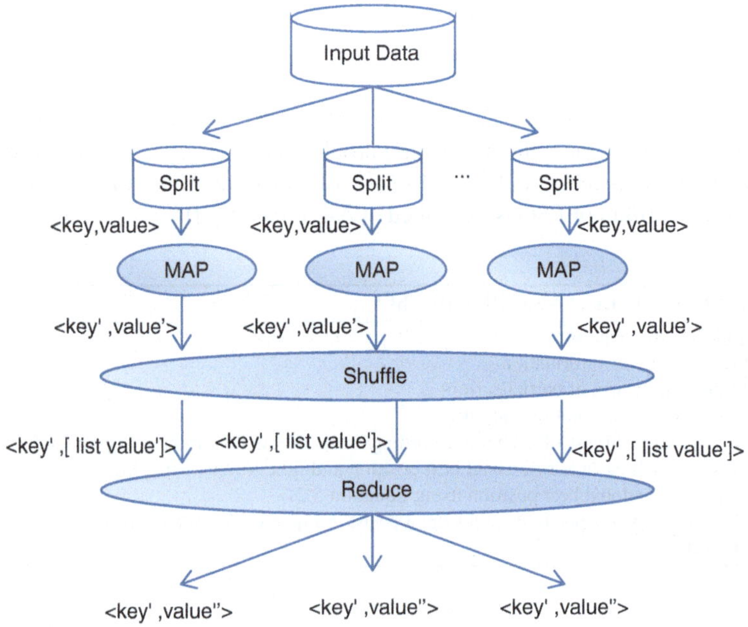

Fig. 5.1 MapReduce flowchart

become widespread in many fields because of its performance, open source nature, installation facilities, and its distributed file system named Hadoop distributed file system (HDFS). In spite of its great popularity, Hadoop MapReduce has significant problems with iterative algorithms.

5.2.3 Apache Spark

Apache Spark, a new framework for Big data processing, designed to be able to solve the Hadoop's shortcomings. It was developed in 2009 in AMPLab in UC Berkeley University, then it became open source in 2010. It is known for its scalability, flexibility, and rapidity [16]. Spark gains its prosperity from its capacity of performing in-memory computations which means that data does not have to be moving from and to the disk instead it is maintained in the memory. In fact, Spark loads the necessary data needed for a specific application, processes and keeps resulted data in memory for further iterations or processes. Therefore, data are read and written only once, rather than multiple times, when iterative processes are required. It also supports streaming processing and this is considered as a strength point regarding Hadoop framework. Spark is based on resilient distributed dataset (RDD), which is considered as a database table that is distributed among the different nodes of the cluster. The RDDs could be created by reading an external data source or by parallelizing a collection of data. Two major operations could be performed on RDDs, namely *transformations* and *actions*. Transformations apply changes on an RDD and generate a new one. Examples of transformations are Map, Sample, GroupbyKey, and ReducebyKey. Actions are operations that require the generation of an output. Examples of actions are Count, Take, and Foreach.

5.3 Related Works

PSO is widely used in cluster analysis since it uses a whole population of possible solutions that collaborate in the purpose of finding better quality clusters [1, 9, 17]. Merwe and Engelbrecht [17] are the first to propose clustering using particle swarm optimization algorithm. In fact, they proposed two methods for data clustering which are the PSO clustering algorithm and the hybrid method which combines both PSO and k-means and then they compared it with k-means algorithm. PSO clustering method is based on the particle swarm optimization algorithm which constitutes a population of particles considered as potential solutions to the clustering problem, these particles contain each k randomly selected initial centroids. This approach is different from that of k-means because it considers a whole swarm of solutions and not only one as in k-means. The particles move in the search space until a global optimal solution is reached. The hybrid method uses the result of k-means algorithm as one of the initial solutions, then randomly generates the rest of the swarm to finally execute the PSO clustering algorithm. The two approaches were compared with the stand-alone k-means algorithm and they showed significantly better performances in convergence and quantization error. Esmin et al. [9] proposed a clustering method based upon the method of Merwe and Engelbrecht [17] with modifications on the regular fitness function. Since the fitness computation doesn't take into account the number of data vectors in each cluster, result's quality was significantly influenced. The number of data objects within each cluster was introduced in the fitness computation which results in good improvements. Ahmadyfard and Modares [1] proposed a new hybrid PSO and K-means clustering algorithm. It starts by executing PSO and switches to k-means when the algorithm reaches the global optimum region. The switch is obtained when the fitness function remains significantly unchanged after several iterations. This combination takes advantage of the strength points of both k-means and PSO and in the same time overcomes their weaknesses. Since PSO shows poor convergence speed near optimum, it is then combined with k-means to speed up the convergence. This combination brings significant improvement compared with the stand-alone PSO and k-means algorithms. Despite the efficiency of the latter discussed methods to deal with the initialization problem using PSO, they are not able to scale with huge volume of data.

To deal with large-scale data, several methods which are based on parallel frameworks have been designed in the literature [2, 4, 13, 20]. Most of these methods use the MapReduce model for data processing. For instance, Zaho el al. [20] have implemented k-means method through MapReduce model. This method first assigns each data point to the nearest cluster prototypes in the map function. The reduce function then updates the new cluster prototypes. Then, this method iterates calling the two functions several times until convergence. Ben HajKacem et al. have proposed fitting k-prototypes using MapReduce model [4] in order to perform the clustering of mixed large-scale data. This method iterates two main steps until convergence: the step of assigning each data point to the nearest cluster center and the step of updating cluster centers. These two steps are implemented in map and reduce phase, respectively. Ludwing has proposed the parallelization of fuzzy c-means clustering using MapReduce model [13]. This method is based on two MapReduce jobs. The first MapReduce job calculates the cluster centers by scanning the input data set. The second MapReduce job also iterates over the data set and calculates the distances to be used to update the membership matrix as well as to calculate the fitness. Although the performance of the latter discussed methods to deal with large-scale data using MapReduce, they do not provide a solution regarding the sensitivity to the selection of the initial cluster centers.

Aljarah and Ludwig [2] proposed MR-CPSO which is, to the best of our knowledge, the only method that processes large-scale data clustering using MapReduce model as well as PSO to ensure better scalability and better clustering quality. In order to perform clustering using PSO, few steps have to be performed which start by the particles initialization, fitness update, centroids update, and finally personal best and global best update. The proposed implementation based on MapReduce suggests three major modules. The first module is a map reduce module responsible for centroids update. The map function receives the particles information as a key value pair where the key is the particle ID and the value represents all the information related to the particle. The map function extracts from each particle the necessary information that enables the update of the centroids and that is done using the position and velocity update formulas (1.3) and (1.4). Then the reduce function combines all updated information into one file and load it into the distributed file system. Once the centroids are updated, the algorithm launches the second module. The second module is a MapReduce module where the map function performs the data assignment step and the reduce function updates the fitness values for each particle. In fact, the map function retrieves the updated particles information and then receives the data chunk to use, then for each particle, data objects are assigned to the closest center. Once done, a key/value pair is generated where the key represents the particle ID and centroid ID where the data is assigned and the value represents the computed distance. Then, the reduce function uses the generated key/value pairs from the previous map function to compute the new fitness values that have to be stored in the distributed file system. The last module is a simple module that merges the outputs resulted from the different previous modules. In addition to that, personal best are updated for each particle as well as a global best. The updated particle's information are stored in the distributed file system to be used for next iterations. Figure 5.2 presents the modules of MR-CPSO.

Table 5.1 summarizes the existing methods.

Although the attested performance of MR-CPSO to perform large-scale data, it has some considerable shortcomings.

- PSO suffers from a low convergence speed close to the global optimum region. In fact, according to [1], particles tend to move slowly when they reach the convergence region. This could be explained by the fact that in PSO, particles tend to follow the global best one and when the algorithm is about to converge, all the particles are almost in the same convergence region. Therefore every

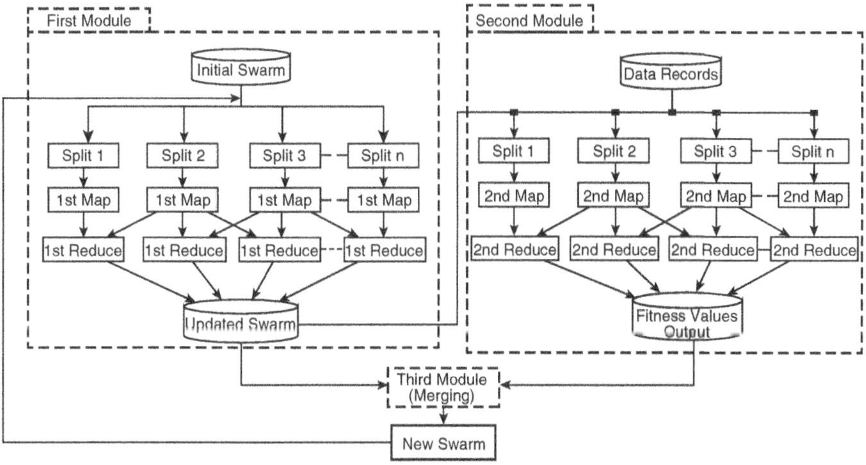

Fig. 5.2 MR-CPSO modules

Table 5.1 Summary of existing methods

	Initialization	Scalability
Merwe and Engelbrecht [17]	+	−
Esmin et al. [9]	+	−
Ahmadyfard and Modares [1]	+	−
Zhao et al. [20]	−	+
Ben HajKacem et al. [4]	−	+
Ludwing [4]	−	+
Aljarah and Ludwig [2]	+	+

particle's $pbestP_i$ and $x_i(t)$ are almost equal to the $gbestP$. Therefore, the particles velocities and positions are not exhibiting significant changes.

- PSO is an iterative algorithm and it requires to perform some iterations for producing optimal results. In contrast, MapReduce has a significant problem with iterative algorithms. As a consequence, the whole data set must be loaded from the file system into the main memory at each iteration. Then, after it is processed, the output must be written to the file system again. Therefore, many of I/O operations like I/O disk occur during each iteration and this decelerates the running time.

5.4 Proposed Approach: S-PSO for Clustering Large-Scale Data

We propose a new efficient PSO clustering method using Spark. The proposed method S-PSO is based on a new strategy that runs k-means algorithm in the latest stages. In fact, k-means is a very fast algorithm so it will accelerate the convergence and it will not affect the final result quality since PSO is very close to the convergence. Furthermore, the proposed method reads the data set only once in contrast to existing MapReduce implementation of PSO clustering. Hence, we aim to exploit in our implementation the flexibility provided by Spark framework, by using in-memory operations that alleviate the consumption time of existing MapReduce solution [2].

S-PSO method is composed of four major steps, namely *Data assignment and fitness computation step*, *Personal and global best update step*, *Position and velocity update step*, and finally *K-means iteration step*. The main process of the proposed method denoted by S-PSO is described in Fig. 5.3.

5.4.1 Data Assignment and Fitness Computation Step

S-PSO starts by setting an initial swarm where it initializes every particle's position, velocity, personal best position, and personal best fitness. The positions are the initial cluster's centroids and they are randomly retrieved from the data. Therefore, each particle once initialized represents a possible solution of the data clustering.

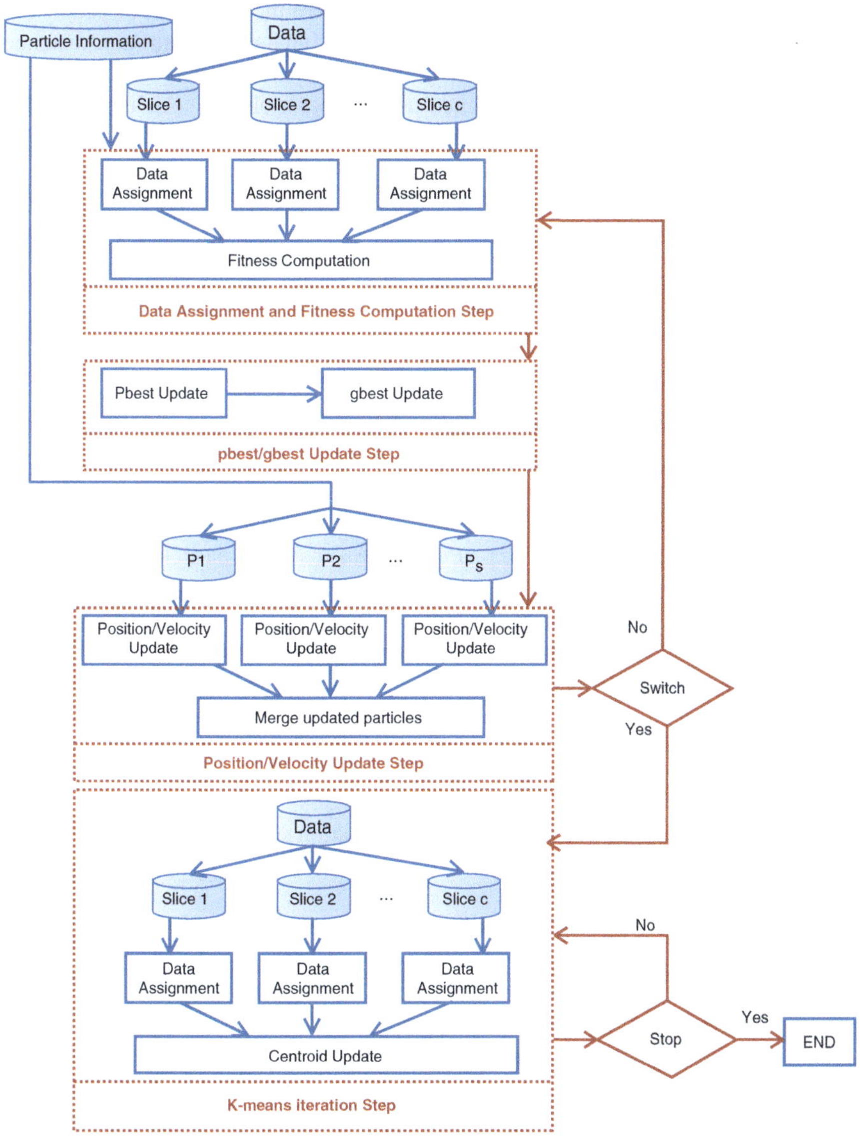

Fig. 5.3 Flowchart S-PSO

This initial swarm encompasses particle's information that will be used for the remaining steps. The data assignment step is a highly expensive operation because it requires to assign the huge amount of data to their closest clusters and this has to be done for every single particle. Since assigning an object is independent from the other objects therefore this step could be performed in parallel. First, the data set is divided into chunks and every chunk is assigned to a map function along with the particle's information. The map function, called *Data assignment*, assigns the data point from its corresponding chunk to the closest cluster in each particle. Then, the map function returns as output a key value pair where the key is composed of the couple particleID and centroidID and the value designates the minimum distance between a data object and the centroidID in a specific particleID.

Once all the data are already assigned to the closest cluster, a reduce function, called *fitness computation* step uses the *reduceByKey()* operation provided by Spark framework to combine the different outputs from the different map functions. The reduce function computes the new fitness value using for that the quantization error given by the following formula:

$$F_i = \frac{\sum_{j=1}^{k}[\sum_{\forall z_p \in C_{ij}} d(z_p, C_j)/|C_{ij}|]}{k} \tag{5.5}$$

where $d(z_p, C_j)$ represents the distance between the data object z_p and the cluster's centroid C_j, $|C_{ij}|$ represents the number of objects assigned to the centroid C_{ij} relative to the particle i, and finally k represents the number of clusters.

Then, the reduce function provides as output a key value pair composed of the particleID as a key and the new fitness value as the value. Let $Z = \{z_1 ... z_n\}$ the input data set. Let $P(t) = \{P_i(t) ... P_S(t)\}$ the set of the particle's information where $P_i(t) = \{x_i(t), v_i(t), pbestP_i(t), pbestF_i(t)\}$ represents the information of particle i in the iteration t where $x_i(t)$ is the position, $v_i(t)$ is the velocity, $pbestP_i(t)$ is the best position, and $pbestF_i(t)$ is the best fitness.

Let $F = \{F_1 ... F_S\}$ the set of fitness values where F_i is the fitness value of the particle i.

Algorithm 2 outlines the data assignment and fitness computation step.

Algorithm 2 Data assignment and fitness computation step

Input: Z: input data set, P(t): particle information
Output: F: fitness values
Split the data set Z into c chunks $Z = \{Z^1 ... Z^c\}$
% Map Phase
Let Z^j be assigned to map j
for each $z_p \in Z^j$ **do**
    for each $P_i(t) \in P(t)$ **do**
        $x_i(t) \leftarrow$ Extract position from $P_i(t)$
        Assign data objects to their closest centroid by computing the euclidean distance
        Let distance the minimum computed distance
        Let CentroidID the index of the centroid where the object z_p is assigned
        Let ParticleID be the index of the particle i
    end for
    Emit (key: ParticleID, CentroidID /value: distance)
end for
% Reduce Phase
for each $P_i(t) \in P$ **do**
    Compute fitness value F_i using Equation 5.5
    Emit (key: ParticleID /value: F_i)
end for

5.4.2 *Pbest and Gbest Update Step*

When the new particle's fitness is computed, it is automatically stored in an RDD distributed among the cluster's nodes. However, since Pbest and gbest update is not an expensive step and it does not require to be performed in parallel that's why the RDD containing the particle's fitness is collected which means it is returned to the driver program to be used in a serial way. Now, each particle updates its personal best position. For the gbest update, the particle having the best fitness value (the smallest quantization error) is identified as gbest particle.

Let $pbestF(t) = \{pbestF_1(t)...pbestF_S(t)\}$ the set of personal best fitness values where $pbestF_i(t)$ is the pbestF of the particle i at iteration t.
Let $pbestP(t) = \{pbestP_1(t)...pbestP_S(t)\}$ the set of personal best position where $pbestP_1(t)$ is the pbestP of the particle i at iteration t. Let $gbestP$ be the position of the best particle.
Algorithm 3 outlines the pbest and gbest update step.

Algorithm 3 Pbest and gbest update step

```
Input: F, pbestF(t), pbestP(t)
Output: pbestF(t+1), pbestP(t+1), gbestP
gbestP ← Ø
for each particle P_i(t) ∈ P(t) do
    pbestF_i(t+1) ← Ø
    pbestP_i(t+1) ← Ø
    if (pbestF_i(t) ≤ F_i) then
        pbestF_i(t+1) ← pbestF_i(t)
        pbestP_i(t+1) ← pbestP_i(t)
    else
        pbestF_i(t+1) ← F_i
        pbestP_i(t+1) ← x_i(t+1)
    end if
end for
Let i* is the index of particle having the best fitness value
gbestP ← x_{i*}(t)
```

5.4.3 *Position and Velocity Update Step*

To take advantage of the parallel environment, S-PSO starts by splitting the particles information among different map functions to perform the velocity and position update using Eqs. (5.3) and (5.4).

Then, the reduce function merges the results provided from the different map functions into one single RDD using for that a *reduceByKey()* operation.

Once finished, the data set and the particle's information stored both in RDDs are persisted in memory for the next iteration and are not returned to the disk. The persistence is performed using the operation *cache()* or *persist()*.

Let $x(t) = \{x_1(t)...x_S(t)\}$ the set of position values where $x_i(t)$ is the position of the particle i at iteration t. Let $v(t) = \{v_1(t)...v_S(t)\}$ the set of velocity values where $v_i(t)$ is the velocity of the particle i at iteration t.

Algorithm 4 outlines the position and velocity update step.

Algorithm 4 Position and velocity update step

Input: $gbest P$, $P(t)$
Output: P(t+1)
% Map Phase
Let $P_i(t)$ be assigned to a map function i
$x_i(t+1) \leftarrow \emptyset$
$v_i(t+1) \leftarrow \emptyset$
Compute the new position value $x_i(t+1)$ using 5.4
Compute the new velocity value $v_i(t+1)$ using 5.3
Emit(key: 1/ value: $P_i(t+1)$)
% Reduce Phase
Merge outputs of the different map functions
Emit ($P(t+1)$)

S-PSO continues iterating until it almost reaches the global optimum region where it becomes very slow. In order to overcome this problem, when the S-PSO reaches the switch condition, it automatically switches to k-means algorithm to take advantage of its speed.

The switch is realized when the variable Time-To-Start is reached: it determines the iteration number where the switch has to occur.

5.4.4 K-Means Iteration Step

When the algorithm switches to k-means, it takes as input the final global best position retrieved from PSO to serve as an initial cluster centroid. K-means is composed of two major steps: data assignment and centroids update. The data assignment step is a map function that takes as input the data chunk and the initial clusters centroids, then it assigns the data objects to the closest cluster. For that, it generates as output a list of key/value pair where key represents the index of the cluster where the data object is assigned and the value represents the data vector.

The centroid update step is a reduce function responsible for merging the different outputs of the map function and for updating the clusters centroids using the mean operation in each cluster. K-means iteration step iterates until it reaches the maximal number of iterations.

Let $C(t) = \{c_1(t)...c_k(t)\}$ the set of cluster centroids at iteration t. Algorithm 5 outlines the k-means iteration step.

Algorithm 5 K-means iteration step

Input: Z, $gbestP$
Output: C(t+1)
Split the data set Z into c chunks $Z = \{Z^1 ... Z^c\}$
C(t) $\leftarrow$ $gbestP$
% Map Phase
Let Z^j assigned to map j
for each $z_p \in Z^j$ **do**
  for each $C_i(t) \in C(t)$ **do**
    Assign data objects to their closest centroid by computing the euclidean distance
    Let CentroidID the index of the centroid where the object z_p is assigned
  end for
  Emit (key:CentroidID /value: z_p)
end for
% Reduce Phase
for each $C_i(t) \in C(t)$ **do**
  Update centroid $C_i(t+1)$
  Emit (key: CentroidID /value: $C_i(t+1)$)
end for

Algorithm 6 outlines the overall steps of S-PSO.

Algorithm 6 S-PSO algorithm

Input: Z: input data set, Iter: maximal iteration number, S: swarm size, k: number of clusters, Time-To-Start: iteration number for switching to K-means, P(t): initial swarm's information
Output: C_f: Final centroids
i $\leftarrow$ 1
switch $\leftarrow$ false
while ((i< Iter) and (!switch)) **do**
  % Data assignment and fitness computation step
  % Pbest and gbest Update step
  % Position and velocity update step
  i++
  if (i =Time-To-Start) **then**
    switch $\leftarrow$ true
    $C = gbestP$
  end if
end while
if (switch =true) **then**
  j $\leftarrow$ 1
  repeat
    % K-means iteration step
    j $\leftarrow$ j + 1
  until $(j = Iter)$
end if

5.5 Theoretical Analysis

5.5.1 Complexity Analysis

Complexity analysis aims to provide the time, space, and I/O complexities of our proposed method S-PSO and compares it to MR-CPSO proposed by [2] since it is the only work dealing with Big data clustering using PSO.

The following notations are used: n the data set size, k the number of clusters, c the number of data chunks, I the number of iterations, s the swarm size, and P the size of the list containing the particle's information.

5.5.1.1 Time Complexity

The most expensive operation in PSO is the data assignment step where each data object has to compute its distance to all the clusters of each particle in the swarm, then this has to be repeated several times. Therefore, the time complexity of PSO could be estimated to $O(n.k.s.I)$.

The S-PSO splits the input data into several chunks that could be processed simultaneously. So instead of processing n data object in every iteration which is the case for PSO, S-PSO will use n/c data item per iteration. Hence, at each iteration S-PSO takes $O(n/c.k.s)$ time. Similar to S-PSO, MR-CPSO works on chunks, thus it takes $O(n/c.k.s)$ time for each iteration. However, regarding the number of iterations, MR-CPSO executes PSO for I times while for S-PSO the number of iteration is divided into I_1 for running PSO and I_2 for running k-means where I_1+I_2=I ($I_1, I_2 < I$). Since K-means is a very fast algorithm which means that executing it for I_2 times will definitely reduce the overall consumed time.

Therefore, the overall time complexity for S-PSO is estimated to $O(n/c.k.s.I_1 + n/c.k.I_2)$. While the overall time complexity for MR-CPSO is estimated to $O(n/c.k.s.I)$. Hence, S-PSO decreases the time complexity of MR-CPSO from $O(n/c.k.s.I)$ to $O(n/c.k.s.I_1 + n/c.k.I_2)$ where $I_1, I_2 < I$.

5.5.1.2 Space Complexity

The S-PSO and MR-CPSO store in the memory a data set of size n where it is used in the data assignment and fitness computation step. In addition to that they initialize and store a list containing the particles information. Therefore, the space complexity of S-PSO and MR-CPSO is estimated to $O(n + P)$.

5.5.1.3 Input/Output Complexity

S-PSO reads the data chunk from disk only once and persists it in the memory. Therefore, the I/O complexity of S-PSO is evaluated by $O(n/c)$. While the MR-

CPSO has to access the disk I times corresponding to the number of iterations. Hence, the I/O cost of MR-CPSO is evaluated by $O(n/c.I)$. As a result, S-PSO can reduce the I/O complexity of MR-CPSO from $O(n/c.I)$ to $O(n/c)$.

5.5.2 *Time-To-Start Variable Analysis*

In PSO, when particles approach the global optimum region, they tend to become very slow. In order to speed up the convergence, k-means known for its speed was integrated in the latest stages of PSO. This way, S-PSO starts by processing PSO first, then it switches to k-means once it approaches the convergence area. We aim by this combination to take advantage of PSO's capacity of finding good quality results on one hand and on the other hand take advantage of k-mean's speed. The switching between the two algorithms is conditioned by a variable that we introduce called Time-To-Start. Time-To-Start designates the iteration number where the switch has to occur. The choice of this variable can influence both time and quality. If this variable is picked up to be in the beginning of PSO, we obtain a low quality result but a very reduced execution time and the opposite is correct. Therefore, this variable has to be chosen in a way that ensures balance between quality and time.

5.6 Experiments and Results

5.6.1 *Methodology*

In order to evaluate the efficiency of S-PSO method, we performed experiments that aim to figure out three major points. (i) How efficient S-PSO method is when applied to large-scale data compared to existing methods? (ii) How the Time-To-Start variable can improve the performance of the proposed method? (iii) How the Spark framework can enhance the scalability of the proposed method when dealing with large-scale data?

5.6.2 *Environment and Data Sets Description*

Experiments were realized using a machine of 16 GB of RAM and $1T$ of disk having 8 cores, it uses Apache Spark version 2.1.1 and scala version 2.1.1 running on a Ubuntu version 16.04. We conducted the experiments on the following data sets:

- Simulated data set: four series of large-scale data sets are generated using the gaussian distribution where the mean is 350 and the sigma is 100. The data sets range from 1 million to 8 millions data points. Each data point is described using

10 attributes. The numeric values are generated with gaussian distribution. In order to simplify the names of the simulated data sets, we use the notations D1M, D2M, D4M, and D8M to denote a simulated data set containing 1, 2, 4, and 8 millions data points, respectively.

- Magic: is a real data set which represents the results of registration simulation of high energy gamma particles in a ground-based atmospheric Cherenkov gamma telescope using the imaging technique. The magic data set contains 19,020 instances having each 10 attributes. The clustering process for this data set identifies whether the energy registered is gamma or not. This data set was obtained from UCI machine learning repository.[1]
- KDD Cup data set (KDD): is a real data set which consists of normal and attack connections simulated in a military network environment. The KDD data set contains about 5 millions connections. Each connection is described using 33 attributes. The clustering process for this data set detects the type of the attacks among all connections. This data set was obtained from UCI machine learning repository. [2]
- Household data set (House) : is a real data set which represents the results of measurements of electric power consumption in household. The House data set contains 2,075,259 data points. Each data point is described using 10 attributes. The clustering process for this data set identifies the types of electric consumption in household. This data set was obtained from UCI machine learning repository.[3]
- CoverType: is a real data set that represents cover type for 30×30 meter cells from US Forest. CoverType data set contains 581,012 instances having each 54 attributes. This data set helps to predict the type of the tree from 7 different types. The real data set is obtained from the UCI[4]. Statistics of these data sets are summarized in Table 5.2.[4]

Table 5.2 Summary of the data sets

Data set	Number of data points	Number of attributes	Domain
D1M	1,000,000	10	Simulated
D2M	2,000,000	10	Simulated
D4M	4,000,000	10	Simulated
D8M	8,000,000	10	Simulated
Magic	19,020	10	Gamma particle's energy
KDD	4,898,431	33	Intrusion detection
House	2,075,259	10	Electricity
CoverType	581,012	54	Agriculture

[1] https://archive.ics.uci.edu/ml/machine-learning-databases/magic/.

[2] https://archive.ics.uci.edu/ml/machine-learning-databases/kddcup99-mld/.

[3] https://archive.ics.uci.edu/ml/machine-learning-databases/00235/.

[4] https://archive.ics.uci.edu/ml/machine-learning-databases/covtype/.

5.6.3 Performance Measures

Evaluating the performance of our proposed method is an important step that takes into account evaluation measures in order to assess the algorithm's quality. The evaluation is addressed towards the approach's scalability and robustness in addition to the quality of clustering results.

- In order to evaluate the scalability, we use the speedup, scaleup, and sizeup measures, which are defined as follows.
  - The speedup measure consists of fixing the data set size and varying the number of computer nodes. Therefore, clustering is performed with one computer node then with m computer nodes, having respectively T_1 and T_m as running time [19]. The speed up is measured as follows:

$$\text{Speedup} = T_1 / T_m \tag{5.6}$$

    The more the speedup is close to linear, the best the algorithm is.
  - While the speedup measure keeps the data set size constant and increases the computer nodes, the scaleup measure increases both the data set size and the number of computer nodes. Therefore, the clustering is performed with one computer node and s as data set size, then it is performed with m nodes with $m * s$ as a data set size [19]. The scaleup measure is given by the following formula:

$$\text{Scaleup} = T_{1s} / T_{ms} \tag{5.7}$$

    The algorithm is considered scalable when T_{1s} is almost equal to T_{m*s}.
  - The sizeup measure holds constant the number of computer nodes and increases m times the size of the data set. It helps to measure the capacity of the algorithm to support m times larger data set [19]. The sizeup measure is given by the following formula:

$$\text{Sizeup} = T_s / T_{ms} \tag{5.8}$$

    An algorithm that has a good sizeup takes m times longer from executing s and $m * s$.

- In order to evaluate the quality of the proposed method, the Running Time and the Quantization Error (QE) given by Eq. (5.5) are used. The QE measures the distance differences between a center and the data objects within that center.

Table 5.3 Comparison of the running time and quantization error of S-PSO versus existing methods

Data set	K-means	PSO	S-PSO
Magic	8	37	25
KDD	4350	10844	2741
House	500	2188	460
CoverType	134	1325	398
Magic	14693	10762	10861
KDD	1.3E12	7.1E11	6.8E11
House	114	53	49
CoverType	1002728	1053373	1061763

5.6.4 Comparison of the Performance of S-PSO Versus Existing Methods

In order to perform the experiments the following parameters are fixed, swarm size as 10 particles, the number of iterations as 50, inertia weight (w) as 0.72, acceleration coefficients $(c1, c2)$ as 1.49, and the number of clusters (k) as 5 for all the data sets except for magic and CoverType is respectively 2 and 7. Table 5.3 reports the performance results of the S-PSO compared to existing methods. Concerning the running time, S-PSO outperforms PSO for all the data sets and also outperforms k-means with house and kdd data sets but for magic and CoverType k-means seems to be faster. This could be explained by the fact that magic and CoverType are considered small data sets and they do not require to be clustered in a parallel manner. Instead, if they are processed in a parallel way it will cost additional time. Concerning the quantization error, PSO and S-PSO provide a better quality regarding k-means while S-PSO keeps almost the same quality as for PSO. This is due to the capacity of PSO and S-PSO using a population of candidate solutions in order to explore the search space.

5.6.5 Evaluation of the Impact of Time-To-Start Variable on the Performance of S-PSO

The purpose is to investigate the impact of the Time-To-Start variable on the performance of S-PSO. Table 5.4 outlines the results. When Time-To-Start variable increases and approaches the final stages of S-PSO, the running time gets more important but the quality gets better as the quantization error decreases. In fact, when the algorithm switches to k-means in its early stages which means when Time-To-Start is small, it takes advantage of the speed of K-means but it is deprived from the capacity of PSO of converging to high quality solutions. In fact the more the Time-To-Start value is small, the fastest the algorithm becomes, the less the quality is.

Table 5.4 Impact of Time-To-Start variable

	Time to start				
Data set	10	20	30	40	50
(a) The impact of Time-to-Start variable on the running time					
Magic	11	17	21	24	28
KDD	421	1416	2077	2789	3206
House	213	383	403	460	661
CoverType	114	229	260	398	493
(b) The impact of Time-to-Start variable on the QE values					
Magic	11301	10944	10630	10294	10365
KDD	9.1E11	9E11	8.9E11	8.9E11	8.9E11
House	65	58	58	50	53
CoverType	1294955	1150217	1139999	1101694	1192713

Therefore, it is important to figure out a compromise between time and quality. In our case, since PSO gets very slow near to convergence the Time-To-Start variable is chosen to be in the last ten iterations. This way switching to k-means will not affect the final result since S-PSO is almost converging and it will reduce the execution time.

5.6.6 Scalability Analysis

Figure 5.4a–d outlines the running time of the proposed method on D1M, D2M, D4M, and D8M, respectively. We can notice that for all data sets the running time decreases as the number of cores increases. For instance, for D4M, the running time decreases from 3987 with 1 core to 730 on 8 cores which means that it decreases over 5 times faster.

To evaluate the speedup for the proposed method, we maintain the constant data set size and we increase the number of cores from 1 to 8. Figure 5.5a–d illustrates the speedup results of respectively D1M, D2M, D4M, and D8M. The overall results show a good speedup for the proposed method. Actually, for all the data sets, when the number of cores goes from 1 to 4, the speedup results are very close to the linear and therefore a very good speedup. When the number of cores exceeds 4, S-PSO's performances start to decrease where the speedup is moving far from the linear. This is due to the additional communication time required to manage the increase of core's number.

Scaleup aims to measure the capacity of an algorithm to maintain the same running time while increasing the data set size with direct proportion to the number of cores. For evaluating the scaleup of our proposed method, we increase both the size of the data set and the number of cores. For investigating the scaleup, we used the D1M, D2M, D4M, and D8M data sets with respectively cores equal to 1, 2, 4, and 8. The results are plotted in Fig. 5.6. The ideal algorithm is the one having its

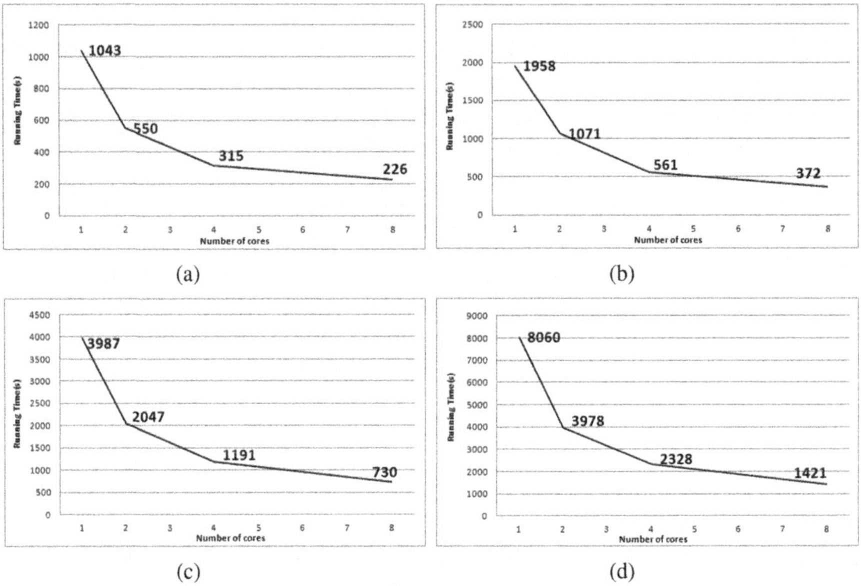

Fig. 5.4 Running time of S-PSO on the different simulated data sets. (**a**) D1M data set. (**b**) D2M data set. (**c**) D4M data set. (**d**) D8M data set

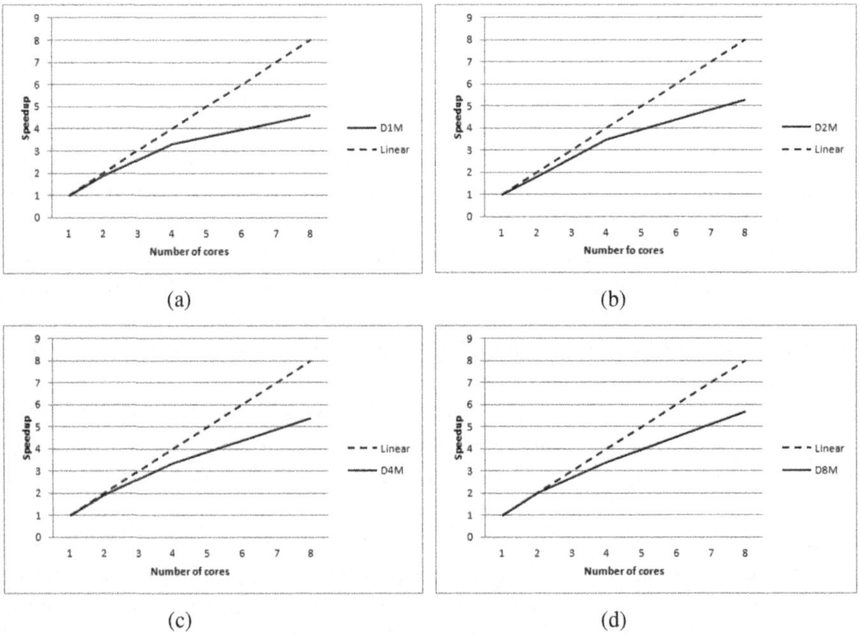

Fig. 5.5 Speedup results of S-PSO on simulated data sets. (**a**) D1M speedup. (**b**) D2M speedup. (**c**) D4M speedup. (**d**) D8M speedup

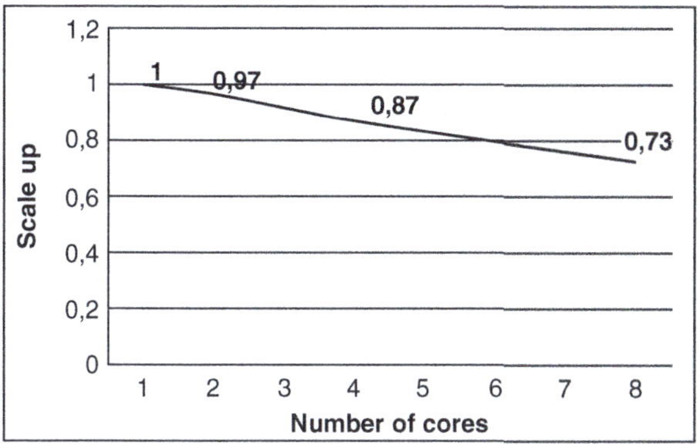

Fig. 5.6 Scaleup results of S-PSO

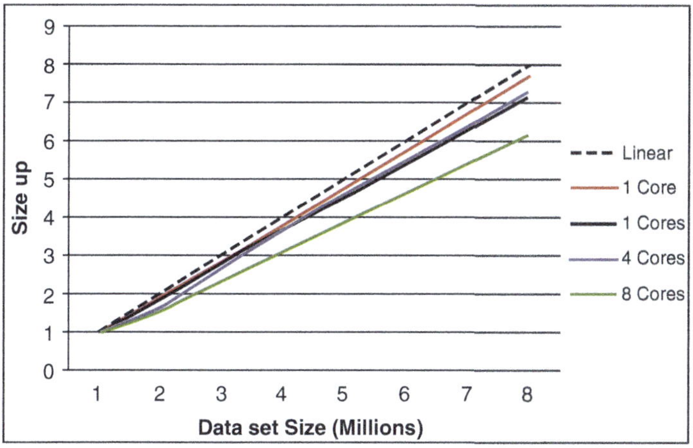

Fig. 5.7 Sizeup results of S-PSO

scaleup values very close or equal to 1. In our case, our proposed method shows a good scaleup. The overall results are approximately similar to D1M, D2M, D4M, and D8M using respectively 1, 2, 4, and 8 cores. Therefore S-PSO scales well with scaleup values ranging between 1 and 0.73.

To evaluate the sizeup of the proposed method S-PSO, the number of cores is kept constant while we increase the data set size in order to evaluate the behavior of S-PSO with the increasing volume of data. Figure 5.7 outlines the obtained sizeup results for respectively 1, 2, 4, and 8 cores. The obtained results show a good sizeup of our proposed method. For instance, for 1 core, the sizeup results are almost equal to the linear reaching 7.7 for D8M.

5.7 Conclusion

Big data clustering is an important field that requires special methods for dealing with high volume of data. Existing methods tried to fit the clustering algorithm based on PSO into MapReduce model. However, due to nonsuitability of MapReduce on iterative algorithms and the low convergence speed of PSO, we proposed a new large-scale data clustering method based on Spark framework and an adapted version of PSO combined with k-means. The proposed method, evaluated on both real and simulated data sets, has shown good results according to quality and scalability measures. As future works, we suggest combining S-PSO with other techniques to automatically looking for the number of clusters k since this parameter must be per-configured in advance. Moreover, we might think of applying feature selection algorithms to select most relevant features to use in S-PSO. This fact can reduce the heavy computation required in each iteration due to the high dimensionality.

References

1. A. Ahmadyfard, H. Modares, Combining PSO and k-means to enhance data clustering, in *International Symposium on Telecommunications, 2008* (2008), pp. 688–691
2. I. Aljarah, S.A. Ludwig, Parallel particle swarm optimization clustering algorithm based on MapReduce methodology, in *2012 Fourth World Congress on Nature and Biologically Inspired Computing (nabic)* (2012), pp. 104–111
3. G.P. Babu, M.N. Murty, Simulated annealing for selecting optimal initial seeds in the k-means algorithm. Indian J. Pure Appl. Math. **25**(1–2), 85–94 (1994)
4. M.A. Ben HajKacem, C.E. Ben N'cir, N. Essoussi, MapReduce-based k-prototypes clustering method for big data, in *Proceedings of Data Science and Advanced Analytics* (2015), pp. 1–7
5. M.E. Celebi, H.A. Kingravi, P.A. Vela, A comparative study of efficient initialization methods for the k-means clustering algorithm. Expert syst. Appl. **40**(1), 200–210 (2013)
6. C.P. Chen, C.-Y. Zhang, Data-intensive applications, challenges, techniques and technologies: a survey on big data. Inf. Sci. **275**, 314–347 (2014)
7. J. Dean, S. Ghemawat, MapReduce: simplified data processing on large clusters. Commun. ACM **51**(1), 107–113 (2008)
8. K.-L. Du, M. Swamy, *Search and Optimization by Metaheuristics: Techniques and Algorithms Inspired by Nature* (Birkhäuser, Basel, 2016)
9. A.A.A. Esmin, D.L. Pereira, F. De Araujo, Study of different approach to clustering data by using the particle swarm optimization algorithm, in *IEEE Congress on Evolutionary Computation, 2008. CEC 2008 (IEEE World Congress on Computational Intelligence)* (2008), pp. 1817–1822
10. A.A. Esmin, R.A. Coelho, S. Matwin, A review on particle swarm optimization algorithm and its variants to clustering high-dimensional data. Artif. Intell. Rev. **44**(1), 23–45 (2015)
11. V. Gorodetsky, Big data: opportunities, challenges and solutions, in *Information and Communication Technologies in Education, Research, and Industrial Applications* (2014), pp. 3–22
12. K. Krishna, M.N. Murty, Genetic k-means algorithm. IEEE Trans. Syst. Man Cybern. B Cybern. **29**(3), 433–439 (1999)
13. S.A. Ludwig, MapReduce-based fuzzy c-means clustering algorithm: implementation and scalability. Int. J. Mach. Learn. Cybern. **6**(6), 923–934 (2015)

14. J. MacQueen et al., Some methods for classification and analysis of multivariate observations, in *Proceedings of the Fifth Berkeley Symposium on Mathematical Statistics and Probability*, vol. 1 (1967), pp. 281–297
15. R. Poli, J. Kennedy, T. Blackwell, Particle swarm optimization. Swarm Intell. **1**(1), 33–57 (2007)
16. R. Shyam, B.G. HB, S. Kumar, P. Poornachandran, K. Soman, Apache spark a big data analytics platform for smart grid. Proc. Technol. **21**, 171–178 (2015)
17. D. Van der Merwe, A.P. Engelbrecht, Data clustering using particle swarm optimization, in *The 2003 Congress on Evolutionary Computation, 2003. CEC'03*, vol. 1 (2003), pp. 215–220
18. D. Xu, Y. Tian, A comprehensive survey of clustering algorithms. Ann. Data Sci. **2**(2), 165–193 (2015)
19. X. Xu, J. Jager, H.-P. Kriegel, A fast parallel clustering algorithm for large spatial databases, in *High Performance Data Mining* (Springer, Berlin, 1999), pp. 263–290
20. W. Zhao, H. Ma, Q. He, Parallel k-means clustering based on MapReduce, in *IEEE International Conference on Cloud Computing* (2009), pp. 674–679

Chapter 6
Data Stream Clustering for Real-Time Anomaly Detection: An Application to Insider Threats

Diana Haidar and Mohamed Medhat Gaber

6.1 Introduction

A data stream is a continuous acquisition of data generated from various source(s) in a dynamic environment, typically in a high velocity, leading to accumulation of large volumes of data. This characterisation leads to a typical Big Data computational problem. The dynamic nature of a data stream imposes a change in the data over time. In real-time data streaming, a change refers to an anomalous data that deviates from the normal baseline (e.g. credit card fraud, network intrusion, cancer, etc.). The ultimate aim of such stream mining problems is to detect anomalous data in real-time.

The absence of prior knowledge (no historical database) is often entangled to a real-time stream mining problem. The anomaly detection system is required to employ unsupervised learning to construct an adaptive model that continuously (1) updates with new acquired data, and (2) detects anomalous data in real-time. The system usually acquires data as segments and identifies the *outliers* in the segment as anomalous. An outlier is an observation that deviates so much from other observations as to arouse suspicion that it was generated by a different mechanism [23]. To detect outliers in a stream mining problem, several approaches have been proposed, nevertheless, unsupervised clustering has been successfully applied to identify the patterns in the data and spot outliers [3]. In this work, we select the insider threat problem as a real-world application to detect malicious insider threats (outliers) in real-time. With the absence of labelled data (no previously logged activities executed by users in an organisation), the insider threat problem poses a challenging stream mining problem.

D. Haidar (✉) · M. M. Gaber
Birmingham City University, Birmingham, UK
e-mail: diana.haidar@bcu.ac.uk; mohamed.gaber@bcu.ac.uk

O. Nasraoui, C.-E. Ben N'Cir (eds.), *Clustering Methods for Big Data Analytics*, Unsupervised and Semi-Supervised Learning,
https://doi.org/10.1007/978-3-319-97864-2_6

Insider threat detection is an emergent concern for academia, industries, and governments due to the growing number of insider incidents in recent years. An insider is a current or former employee, contractor, or business partner of an organisation who has authorised access to the network, system, or data (e.g. trade secrets, organisation plans, and intellectual property) [35]. Malicious insider threats are attributed to insiders who exploit their privileges with the intention to compromise the confidentiality, integrity, or availability of the system or data. According to the 2011 Cybersecurity Watch Survey [10], 21% of attacks are attributed to insiders in 2011, while 58% are attributed to outsiders. However, 33% of the respondents inspect the insider attacks to be more costly, compared to 51% in 2010. For instance, Harold Martin, a former top-security contractor at the National Security Agency (NSA), was recently convicted for stealing around 500 million pages of national defence information over the course of 20 years [38]. Earlier in 2013, Snowden's attack was reported as the biggest intelligence leakage in the USA [42]. Edward Snowden, a former contractor to the NSA, disclosed 1.7 million classified documents to the mass media.

The challenge of the insider threat detection problem lies in the variety of malicious insider threats in the data sets. Each malicious insider threat is devised of a complex pattern of anomalous behaviours carried out by a malicious insider, thus making it difficult to detect **all** anomalous behaviours per threat. Analytically, some anomalous instances (behaviours) which exist in a dense area of normal instances have a high similarity to normal instances. These anomalous instances are difficult to detect and may be missed by the detection system.

Based on the challenge of the problem, we formulate this work with the aim to detect **any-behaviour-all-threat**; it is sufficient to detect **any** anomalous behaviour in **all** malicious insider threats. In other words, we can hunt a malicious insider threat by at least detecting one anomalous behaviour among the anomalous behaviours associated with this threat. We call this approach *threat hunting*. The design of the proposed approach with such a relaxing condition contributes in reducing the frequent false alarms.

Figure 6.1 illustrates a continuous data stream of behaviours (instances) including normal behaviours and anomalous behaviours. Each arrow denotes a behaviour (instance) X^t executed by a user at a specific period of time. Let the blue arrows represent the normal behaviours. Let the green arrows and the red arrows represent

Fig. 6.1 Continuous data stream of behaviours

the anomalous behaviours which belong to the malicious insider threats T_1 and T_2, respectively. To detect a malicious insider threat T_1, it is required to detect **any** green behaviour X^t. Hence, it is essential to detect **any** of the anomalous behaviours per malicious insider threat; the aim to detect any-behaviour-all-threat. Note that the proposed approach may detect **more than one** behaviour which belong to a malicious insider threat, nevertheless, the ultimate aim is to detect one anomalous behaviour per threat.

Based on the above argument, the feature space is defined as a set of features which describe the user's behaviour. Each feature is extracted from the data set logs to represent a user's behaviour related to a particular activity. A feature vector (i.e. instance, behaviour) represents a set of feature values (i.e. actions, commands) evaluated at a period of time. A more detailed description about the feature space is provided in Sect. 6.3.

Several machine learning approaches have been suggested to address the insider threat problem. However, these approaches still suffer from a high number of false alarms. A recent real-time anomaly detection system, named RADISH, based on k nearest neighbours (k-NN) is proposed by Bose et al. [5]. The experimental results showed that 92% of the alarms flagged for malicious behaviour are actually benign [false positives (FPs)].

To address the shortcoming of the high number of false alarms, we propose a streaming anomaly detection approach, namely Ensemble of Random subspace Anomaly detectors In Data Streams (E-RAIDS). We presented a preliminary version of E-RAIDS (coined RandSubOut) in [22]. E-RAIDS is built on top of the established outlier detection techniques [Micro-cluster-based Continuous Outlier Detection (*MCOD*) or Anytime Outlier Detection (*AnyOut*)] which employ clustering over continuous data streams. The merit of E-RAIDS is its capability to detect malicious insider threats in real-time (*based on the definition of real-time in terms of window iterations as discussed in Sect.* 6.4).

E-RAIDS learns an ensemble of p outlier detection techniques (either MCOD or AnyOut), such that each model of the p models learns on a random feature subspace. The acquired data is accumulated in a temporary *buffer* of pre-defined capacity (equals to a fixed window size). So that, at each window iteration, each of the p models in the ensemble is updated with the acquired data, and local outliers are identified in the corresponding feature subspace.

Let $p = n + 1$ denote the number of feature subspaces selected randomly, such that n is set to the number of features (dimension) of the feature space F in the community data set. n represents the number of feature pairs (i.e. two features per subspace), and 1 represents the whole feature space (i.e. all features). In this way, E-RAIDS is capable to detect *local outliers* in the n feature pairs, as well as *global outliers* in the whole feature space. Hence, E-RAIDS employs the idea of random feature subspaces to detect local outlier(s) (anomalous behaviour(s)) which might not be detected over the whole feature space. These anomalous behaviour(s) might refer to malicious insider threat(s).

E-RAIDS introduces two factors: (1) a survival factor vf_s, which confirms whether a subspace votes for an alarm, if outlier(s) survive for a vf_s number of window iterations; and (2) a vote factor vf_e, which confirms whether an alarm should be flagged, if a vf_e number of subspaces in the ensemble vote for an alarm. E-RAIDS employs an aggregate component that aggregates the results from the p feature subspaces, in order to decide whether to generate an alarm. The rationale behind this is to reduce the number of false alarms.

The main contributions of this chapter are summarised as follows:

- an ensemble approach on random feature subspaces to detect local outliers (malicious insider threats), which would not be detected over the whole feature space;
- a survival factor that confirms whether outlier(s) on a feature subspace survive for a number of window iterations, to control the vote of a feature subspace;
- an aggregate component with a vote factor to confirm whether to generate an alarm, to address the shortcoming of high number of FPs;
- a thorough performance evaluation of E-RAIDS-MCOD and E-RAIDS-AnyOut, validating the effectiveness of voting feature subspaces, and the capability to detect (more than one)-behaviour-all-threat detection (Hypothesis 2) in real-time (Hypothesis 1).

E-RAIDS extends the preliminary version RandSubOut [22], where it upgrades the selection of the set of outliers (anomalous behaviours) identified at a window iteration over a feature subspace to improve the detection performance of malicious insider threats over the whole ensemble. This is later described in Sect. 6.3.3.1. Unlike RandSubOut, we evaluate E-RAIDS on community data sets such that each community is richer with malicious insider threats which map to a variety of scenarios. Moreover, E-RAIDS is evaluated, not only in terms of the number of detected threats and FP Alarms, however, in terms of (1) F1 measure, (2) voting feature subspaces, (3) real-time anomaly detection, and (4) the detection of (more than One)-behaviour-all-threat.

The rest of this chapter is organised as follows. Section 6.2 reviews the techniques which utilised clustering to detect outliers, and the stream mining approaches proposed for insider threat detection. Section 6.3 presents the proposed streaming anomaly detection approach, namely E-RAIDS, for insider threat detection. Experiments and results are discussed in Sect. 6.4. Finally, Sect. 6.5 summarises the chapter and suggests future work.

6.2 Related Work

The absence of labelled data (i.e. low data maturity) reveals that an organisation has no previously logged activities executed by users (no historical database). We address the absence of prior knowledge using unsupervised streaming anomaly detection built on top of established outlier detection techniques.

Clustering has been successfully applied to identify the patterns of the data for outlier detection [3]. The continuous acquisition of data generated from various sources defines the streaming environment of the insider threat problem. Several clustering methods have been proposed to handle the streaming environment. The state of the art presents two primitive clustering methods: Balanced Iterative Reducing and Clustering using Hierarchies (BIRCH) [46], and CluStream [1].

BIRCH is an incremental hierarchical clustering method which was first proposed for very large data sets. BIRCH incrementally and dynamically clusters acquired data instances. It maintains a tree of cluster features (information about clusters) which is updated in an iterative fashion [21]. Thereafter, BIRCH was applied to data stream clustering.

BIRCH was the first to introduce the concepts of micro- and macro-clusters [26]. CluStream is a data stream clustering method that employed those two concepts for the clustering process: online micro-clustering and offline macro-clustering. In the online phase, CluStream scans the acquired data instances and creates micro-clusters in a single pass to handle the big (unbounded) data stream. In the offline phase, CluStream only utilises the micro-clusters and re-clusters into macro-clusters [21].

From BIRCH to CluStream, the concept of data stream clustering is applied, despite the fact that BIRCH includes incremental processing of data instances. *Incremental clustering* processes one data instance at a time and maintains a predefined data structure (i.e. model) that is incrementally updated without the need for model reconstruction [40]. In fact, many incremental methods predate the data stream mining methods. The intrinsic nature of data streams requires methods which implement incremental processing of data instances, in order to handle the resource limitations (i.e. time and memory) [40]. But unlike earlier incremental clustering methods, *data stream clustering* methods require a more efficient time complexity to cope with high data rates [28]. Indeed, the literature stresses the importance of considering the inherent time element in data streams [40]. For instance, a typical data stream clustering method exhibits the temporal aspect of cluster tracking [40]. The dynamic behaviour of cluster over a data stream manifests in the evolving of existing clusters over time; the emergence of new clusters; and the removal of clusters based on a time stamp and/or size. Those cluster updates must be performed on the data structure (i.e. model) very efficiently [28]. Hence, the data stream clustering methods are not only incremental, but are also fast in terms of the inherent temporal aspects.

In this work, data stream clustering is used to underpin the outlier detection techniques for real-time anomaly detection. In the insider threat problem, the temporal aspect is substantial to consider, in order to detect malicious insider threats in real-time (based on the definition of real-time in terms of window iterations as discussed in Sect. 6.4). Hence, data stream clustering methods are more adequate to be utilised in this work than typical incremental clustering methods.

In the following we review the techniques which utilised clustering to detect outliers. Thereafter, we shed light on two outlier detection techniques (MCOD and AnyOut) which employ *data stream clustering*. Those two techniques are later utilised in the proposed framework.

We also review the streaming anomaly detection approaches proposed for insider threat detection.

6.2.1 Clustering for Outlier Detection

It is important to distinguish our objective—to optimise outlier detection—from that of a benchmark of clustering methods developed to optimise clustering (such as DBSCAN [11] and BIRCH [46]). The outliers in these methods [11, 46] are referred to as noise, where they are usually tolerated or ignored. However, in our work, we refer to outliers as anomalous (suspicious) behaviour(s) which may correlate to malicious insider threats. Therefore, clustering is utilised to optimise the detection of outliers.

On the other hand, a benchmark of clustering methods to optimise outlier detection (such as LOF [6] and CBLOF [24]) are developed. Breunig et al. [6] introduced the concept of Local Outlier Factor (LOF) to determine the degree of outlierness of an instance using density-based clustering. The locality of instance (local density) is estimated by the distance to its k nearest neighbours. Thus, the LOF of an instance located in a *dense* cluster is close to 1, while lower LOF is assigned for other instances. He et al. [24] present a measure, called Cluster-based Local Outlier Factor (CBLOF), to identify the physical significance of an outlier using similarity function. The CBLOF of an instance is determined by (1) the distance to its nearest cluster (if it belongs to a small cluster), or (2) the distance to its cluster (if it belongs to a large cluster).

However, the aforementioned methods do not tackle outlier detection in *continuous* data streams. MCOD [28] and AnyOut [3] are two established outlier detection techniques, which employ data stream clustering in *continuous* data streams. MCOD and AnyOut are utilised as building block clustering techniques for the proposed E-RAIDS approach. A detailed description and formalisation for these techniques is found in Sect. 6.3.2.

6.2.2 Streaming Anomaly Detection for Insider Threat Detection

Few approaches have utilised streaming anomaly detection to detect insider threats [36, 37, 41, 45] with no prior knowledge. We give a brief description for these approaches in the following.

Among the emerging interest in deep learning, Tuor et al. [41] present a preliminary effort to utilise deep learning in an online unsupervised approach to detect anomalous network activity in real-time. The authors presented two models: a deep neural network (DNN) model which is trained on each acquired instance only once, and a recurrent neural network (RNN) which learns an RNN per user such that the weights are shared among all users, however, the hidden states are trained per user.

Zargar et al. [45] introduce a Zero-Knowledge Anomaly-Based Behavioural Analysis Method, namely XABA, that learns each user's behaviour from raw logs and network traffic in real-time. XABA is implemented on a big-stream platform without pre-defined or pre-processed activity logs, to handle high rates of network sessions. The authors indicated that XABA reports a low number of FPs, when evaluated on a real traitor scenario.

One of the remarkable approaches is an ensemble of one class SVM (ocSVM), namely Ensemble-based Insider Threat (EIT), proposed by Parveen et al. [36]. The authors proposed an ensemble approach, based on static ocSVM learners, to model a continuous data stream of data chunks (i.e. daily logs). The EIT maintains an ensemble of a k pre-defined number of models M, where the ensemble is continuously updated at each day session upon the learning of a new model M_s. The EIT selects the best $k - 1$ models from the k models, having the minimum prediction error over the data chunk C_s, and appends the new model M_s. The results show that ensemble-ocSVM outperforms ocSVM, where it reports a higher accuracy and almost half the number of FP. The authors extended their work in a future paper [37], where the ensemble approach is applied to unsupervised graph-based anomaly detection (GBAD). The results show that ensemble-ocSVM outperforms the ensemble-GBAD in terms of FP.

The above approaches have shown merit in addressing the insider threat detection problem, however, as aforementioned, they do suffer from high false alarms. In this book chapter, we utilise data stream clustering to detect outliers (malicious insider threats), while reducing the number of false alarms.

6.3 Anomaly Detection in Data Streams for Insider Threat Detection

This section identifies the feature space in the insider threat problem and the categories of the feature set extracted. It then describes and formalises established continuous distance-based outlier detection techniques (MCOD and AnyOut). It presents the proposed *E-RAIDS* for insider threat detection with a detailed description of the feature subspace anomaly detection and the aggregate component (ensemble voting).

6.3.1 Insider Threat Feature Space

In this book chapter, we utilise the synthetic data sets, including a variety of malicious insider threat scenarios, generated by Carnegie Mellon University (CMU-CERT) [16]. The CMU-CERT data sets comprise system and network logs for the activities carried out by users in an organisation over 18 months (e.g. logons, connecting removable devices, copying files, browsing websites, sending emails, etc.). We extract a feature set from these logs in the CMU-CERT data sets according to the literature [5, 30, 31]. This feature set allows us to assess the behaviour of users, and compare it to the previous behaviour of the users or their community of users. We extract each feature considering the evidence it would reveal about an undergoing anomalous behaviour. For example, consider the feature *logon after hours*; this feature if its values is greater than 1, it reveals an evidence of an unusual activity undergoing after the official working hours. Thus, it contributes in the overall decision of the system whether a malicious alarm should be flagged or not.

In the following, we give a more detailed description of the feature set used in this work. We categorise the features into five groups with some examples of each.

- Frequency-based features: assess the frequency of an activity carried out by a community of users over each session slot ($integer$, e.g. *frequency of logon, frequency of connecting devices*);
- Time-based features: assess an activity carried out within the non-working hours period of time ($integer$, e.g. *logon after work hours, device usage after work hours*);
- Boolean features: assess the presence/absence of an activity-related information ($flag = \{0, 1\}$, e.g. *non-empty email-bcc, a non-employee email recipient, sensitive file extension*);
- Attribute-based features: are more specialised features, which assess an activity with respect to a particular attribute (feature) value ($integer$, e.g. *browsing a particular URL (job websites, WikiLeaks)*); and
- Other features: assess the count of other activity-related information. ($integer$, e.g. *number of email recipients, number of attachments to emails*).

This feature set defines the feature space of the insider threat problem, and is used to construct community behaviour profiles for users having the same role (e.g. Salesman, IT admin). A community behaviour profile represents instances (i.e. vectors of feature values) evaluated over session slots, where a session slot represents a period of time from *start time* to *end time*. Each vector of feature values is extracted from the behaviour logs of the community users during a session slot.

In this work, we define the session slot per 4 h to find local anomalous behaviour within a day which would not be detected per day. The rationale behind choosing the session slot *per 4 h* is that this period of time is long enough to extract an instance (i.e. vector of feature values) which provides an adequate evidence of anomalous behaviour. Thus, it supports the system to capture the anomalous behaviours in feature space. If the session slot is chosen per minutes, for example, the extracted instances would lack the adequate evidence of the occurrence of anomalous behaviour. On the other hand, if the session slot is chosen per days/weeks, for example, the period of time will be too long to capture the anomalous behaviour blurred among the normal behaviour in the extracted vector of feature values.

After constructing the community behaviour profiles, we normalise each vector of feature values (over a session slot) to the range [0, 1], and associate it with a class label $\{Normal, Anomalous\}$.

6.3.2 *Background on Distance-Based Outlier Detection Techniques*

The state of the art presents one of the most widely employed techniques for anomaly detection, which are *distance-based* outlier detection techniques. According to the definition [27], an instance X^t is an outlier, if there exists less than k number of neighbours located at a distance at most r from X^t.

In this work, we are interested in *continuous outlier detection* over a data stream, where recent instances arrive and previous instances expire. The demo paper [15] gives a comparison of four continuous distance-based outlier detection techniques: STream OutlieRMiner (STORM) [2]; Abstract-C [44]; Continuous Outlier Detection (COD) [28]; and Micro-cluster-based Continuous Outlier Detection (MCOD) [28]. COD and MCOD have $O(n)$ space requirements, and they have a faster running time than the exact algorithms of both [2] and [44]. Note that since all algorithms are exact, they output the same outliers [15]. According to [15], COD and MCOD demonstrate a more efficient performance compared to STORM and Abstract-C in terms of lower space and time requirements. Hence, the latter are excluded, but not COD and MCOD. Furthermore, based on the experimental evaluation in [28], MCOD outperforms COD over benchmark tested data sets. Hence, MCOD is selected to be utilised as a base learner in the proposed E-RAIDS approach.

The state of the art presents a further continuous distance-based outlier detection technique, called Anytime Outlier Detection (AnyOut) [3]. However, AnyOut has not been compared to the four techniques in the demo paper. We select MCOD and AnyOut as base learners in E-RAIDS to compare their performance. It is worth to note that the aforementioned distance-based outlier detection techniques are implemented by the authors of [15] in the open-source tool for Massive Online Analysis (MOA) [33].

In the following, a brief description of MCOD and AnyOut techniques is provided.

6.3.2.1 Micro-Cluster-Based Continuous Outlier Detection

Micro-cluster-based Continuous Outlier Detection (MCOD), an extension to Continuous Outlier Detection (COD), stems from the adoption of an event-based technique. The distinctive characteristic of MCOD is that it introduces the concept of evolving micro-clusters to mitigate the need to evaluate the range query for each acquired instance X^t with respect to all the preceding active instances. Instead, it evaluates the range queries with respect to the (fewer) centres of the micro-clusters. The micro-clusters are defined as the regions that contain inliers exclusively (with no overlapping). The micro-clustering is fully performed online [28].

Given that, the centre mc_i of a micro-cluster MC_i may or may not be an actual instance X^t; the radius of MC_i is set to $\frac{r}{2}$, such that r is the distance parameter for outlier detection; and the minimum capacity (size) of MC_i is $k+1$. Below we give a brief formalisation of MCOD.

Let k represent the number of neighbours parameter. Let PD represent the set of instances that doesn't belong to any micro-cluster.

The micro-clusters (i.e. centres of micro-clusters) in MCOD are determined as described in [1]. In the initialisation step, seeds (with a random initial value) are sampled with a probability proportional to the number of instances in a given micro-cluster. The corresponding seed represents the centroid of that micro-cluster. In later iterations, the centre mc_i is the weighted centroid of the micro-cluster MC_i.

- Step 1: For each acquired instance X^t, MCOD finds (1) the nearest micro-cluster MC_i, whose centre mc_i is the nearest to it, and (2) the set of micro-cluster(s) R, whose centres are within a distance $\frac{3\times r}{2}$ from their centres.
- Step 2: If $dist(X^t, mc_i) \leq \frac{r}{2}$; such that mc_i is the centre of the nearest cluster MC_i, X^t is assigned to the corresponding micro-cluster.
- Step 3: Otherwise, a range query q for X^t is evaluated with respect to the instances in (1) the set PD and (2) the micro-clusters of the set R.
- Step 4: Consider n: the number of neighbours $N^{t'} \in P$ to X^t, such that $dist(X^t, N^{t'}) \geq \frac{r}{2}$. If $n > \theta k; \theta \geq 1$, then a new micro-cluster with centre X^t is created and the n neighbours are assigned to this micro-cluster.
- Step 5: A micro-cluster whose size decreases below $k+1$ is deleted and a range query similar to that described for X^t is performed for each of its former instances.
- Step 6: An instance X^t is flagged as an outlier, if there exists less than k instances in either PD or R.

6.3.2.2 Anytime Outlier Detection

Anytime Outlier Detection (AnyOut) is a cluster-based technique that utilises the structure of ClusTree [29], an extension to R-tree [20, 39], to compute an outlier score. The tree structure of AnyOut suggests a hierarchy of clusters, such that the clusters at the upper level subsume the fine grained information at the lower level. ClusTree is traversed in top-down manner to compute the outlier score of an acquired instance X^t until it is interrupted by the subsequent (next) instance X^t. Thus, the descent down the tree improves the certainty of the outlier score, nevertheless, the later the arrival of the subsequent instance the higher the certainty [3].

Given that, a cluster is represented by a Cluster Feature tuple $CF = (n, LS, SS)$, such that n is the number of instances in the cluster, LS and SS are respectively the linear sum and the squared sum of these instances. The compact structure of the tree using CF tuples reduces space requirements. From BIRCH [46] to CluStream [1], cluster features and variations have been successfully used for online summarisation of data, with a possible subsequent offline process for global clustering.

Let e_s represent a cluster node entry in ClusTree. Given defined two scores to compute the degree of outlierness of an instance X^t: (1) a mean outlier score is $s_m(X^t) = dist(X^t, \mu(e_s))$, such that $\mu(e_s)$ is the mean of the cluster node entry $e_s \in$ ClusTree where X^t is interrupted; and (2) a density outlier score is $s_d(X^t) = 1 - g(X^t, e_s)$, such that $g(x_i, e_s)$ is the Gaussian probability density of X^t for $\mu(e_s)$ as defined in [3]. Below we give a brief formalisation of AnyOut.

In the case of a constant data stream:

- Step 1: Initialisation: Each X^t in the data stream is assigned with an actual confidence value $conf(X^t) = e^{s(X^t)}$.
- Step 2: Distribute the computation time for each X^t based on the scattered confidences.
- Step 3: For each acquired instance X^t, AnyOut traverses the tree in a top-down manner until the arrival of the instance X^{t+1} in the data stream.
- Step 4: At the moment of interruption, X^t is inserted to the cluster node entry $e \in$ ClusTree, where X^t arrives (pauses).
- Step 5: The outlier score of X^t, according to the specified parameter $s_m(X^t)$ or $s_d(X^t)$, is computed with respect cluster node entry e_s.
- Step 6: The expected confidence value for the outlier score of X^t is updated based on the computation time. *The confidence value (certainty) increases as the computation time increases.*

6.3.3 E-RAIDS Approach

The established continuous distance-based outlier detection techniques (MCOD and AnyOut), described and formalised in Sect. 6.3.2, are utilised as building block data stream clustering techniques for the proposed E-RAIDS approach.

In this work, we propose a streaming anomaly detection approach, namely Ensemble of Random subspace Anomaly detectors In Data Streams (E-RAIDS), for insider threat detection. In other words, E-RAIDS is an ensemble of an established distance-based outlier detection technique (MCOD or AnyOut), such that each model of the ensemble learns on a random feature subspace. The idea of E-RAIDS is to employ an outlier detection technique on a feature subspace, to detect *local outliers* which might not be detected over the whole feature space. These local outliers may refer to anomalous behaviours (instances) attributed to a malicious insider threat. Hence, the ultimate aim of the E-RAIDS approach is to detect any-behaviour-all-threat (*threat hunting* as defined in Sect. 6.1); a process that leads to a reduction of the number of false alarms.

Figure 6.2 presents the E-RAIDS framework. The set of blue arrows represent a data stream, where each arrow represents an instance (feature vector) X^t acquired at a session slot t. Consider the formalisations below:

- window: a segment of a fixed size w that slides along the instances in a data stream with respect to time (i.e. session slots);

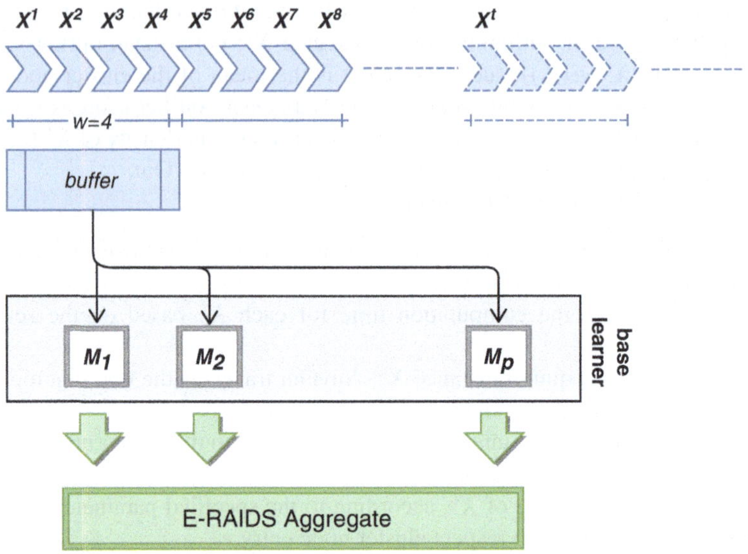

Fig. 6.2 E-RAIDS framework

- *buffer*: a temporary short memory of allocated capacity (equals to w). It temporarily accumulates the instances in a data stream. The *buffer* starts to accumulate the instances in a data stream at the start of the window and stops once the *buffer* is full after w number of instances. The full *buffer* is then input to the base learner component to be processed; and
- window iteration $wIter$: an iteration of the window slide. It starts at the already processed instances (in the previous *buffer*) $procInst$ and ends at $procInst + w$. For instance, $wIter = 0$ starts at $procInst = 0$ and ends at w. $wIter = 1$ starts at $procInst = w$ and ends at $2 \times w$. The window iteration $wIter$ allows the synchronisation between the window slide and the *buffer* accumulation.

Based on the aforementioned formalisations, at each window iteration $wIter$, the *buffer* accumulates w number of instances. Once the *buffer* is full, the instances in the *buffer* are input to the base learner component. A base learner component refers to the distance-based outlier detection technique to be utilised (MCOD or AnyOut). It employs a p number of base models to learn on randomly selected p feature subspaces. A feature subspace $FS_i \subseteq F$ is defined as a subset of features selected from the whole feature space f, where $F = \{f_1, f_2, \ldots, f_n\}$. The rationale behind the idea of random feature subspaces is to detect local outlier(s) (anomalous behaviour(s)) which might not be detected over the whole feature space.

Let $p = n + 1$ represent the number of feature subspaces selected randomly, such that n is set to the number of features (dimension) of the feature space F in the community data set. n represents the number of feature pairs (i.e. 2 features per subspace), and 1 represents the whole feature space (i.e. all features). The p feature subspaces are utilised to build the ensemble of p models $\{M_1, M_2, \ldots, M_n, M_p\}$, such that $\{M_1, M_2, \ldots, M_n\}$ learn on feature pairs, and M_p learns on the whole feature space. In this way, E-RAIDS is capable to detect *local outliers* in the n feature pairs, as well as *global outliers* in the whole feature space.

6.3.3.1 Feature Subspace Anomaly Detection

For each model M_i on a feature subspace FS_i, we define the following data repositories and a survival factor:

- $outSet$: a temporary set of the outliers detected by a base learner (MCOD or AnyOut) at $wIter$;
- $outTempList$: a list that stores the triples (1) an outlier $out \in outSet$, (2) the $wIter$ where it was first detected, and (3) a temporal count $tempC$ which counts the number of subsequent windows out was detected at. It has the form $\langle out, wIter, tempC\rangle$; and
- $subVoteList$: a list that stores the triples (1) a $wIter$, (2) a $subVote$ parameter set to 1 if the feature FS_i votes for an alarm to be generated at $wIter$ and 0 otherwise, and (3) an outlier set $subOutSet$. It has the form $\langle wIter, subVote, subOutSet\rangle$.
- vf_s: a survival factor that confirms whether a feature subspace FS_i votes for an alarm to be generated at $wIter$. In other words, if an outlier out survived (has been detected) for a vf_s number of subsequent windows, then out is defined as a *persistent* outlier. A persistent outlier boosts the chance of an alarm to be generated at $wIter$.

Over each feature subspace FS_i, the base learner (MCOD or AnyOut) processes the *buffer* at the current $wIter$ to update the model M_i. M_i identifies the outlier set $outSet$, which includes (1) the outliers from the *buffer* at the current $wIter$; and (2) the outliers from the previously learned instances before the model being updated. The type (2) outlier refers to an instance that was identified as an inlier, however, turned into an outlier in the current $wIter$.

For each outlier $out \in outSet$, if out does not exist in $outTempList$, a new triple $\langle out, wIter, 1\rangle$ is appended to $outTempList$. In this case, $tempC$ is assigned to 1 to declare that it is the first time an outlier out detected. If out exists in the $outTempList$, $tempC$ is incremented by 1 in the triple for out.

For each outlier $out \in outTempList$, E-RAIDS checks (1) if $tempC = vf_s$, then out survived for a vf_s number of subsequent windows. We call it *persistent outlier*. Thus, a persistent outlier confirms that the FS_i votes for an alarm at $wIter$. Thus, $subVote$ is set to 1 in the triple for $wIter$ in $subVoteList$. (2) If $wIter - tempC = vf_s$, then out has turned into an inlier. We call it *expired outlier*. Thus, the triple for the expired outlier out is removed from the $outTempList$. (3) If $wIter - tempC < vf_s$, then out is neither a persistent outlier nor an expired outlier. We call it *potential outlier*.

Algorithm 1 Feature subspace

```
wIter ← 0
foreach X^t ∈ stream do
    accumulate X^t to buffer
    if buffer is full then
        outSet ← get outliers detected by base
        foreach out ∈ outSet do
            if outinoutTempList then
                a new triple ⟨out, wIter, 1⟩ is appended to outTempList
            else
                increment tempC by 1 for out in outTempList
            end
        end
        foreach out ∈ outTempList do
            if tempC = vf_s then
                set subVote to 1 in subVoteList for the current wIter
                remove out from outTempList
                subOutSet ← persistent out∪ potential outliers
                append subOutSet to subVoteList at wIter
            end
            if wIter − tempC = vf_s then
                remove out from outTempList
            end
        end
        increment wIter
    end
    empty buffer
end
return subVoteList
```

Eventually, $outTempList$ consists of persistent outliers and potential outliers (expired outliers have been removed). E-RAIDS appends persistent outliers and potential outliers in $outTempList$ to the $subOutSet$ in the triple for $wIter$ in $subVoteList$.

In the preliminary version RandSubOut [22], only the persistent outliers in $outTempList$ are appended to the $subOutSet$ in the triple for $wIter$ in $subVoteList$. The $subOutSet$ at $wIter$ represents the set of anomalous behaviours detected over a feature subspace FS_i. As later described in Sect. 6.3.3.2, if the ensemble votes to generate an alarm, then the $subOutSet$ for each feature subspace is utilised to evaluate whether all malicious insider threats are detected (i.e. the aim of any-behaviour-all-threat). The experiments carried out on both versions (E-RAIDS and RandSubOut) showed that E-RAIDS outperforms the latter in terms of detecting more malicious insider threats. Analytically, unlike RandSubOut, E-RAIDS considers further the potential outliers in the $subOutSet$ to check for detected malicious insider threats. Thus, the use of potential outliers significantly boosts the detection performance of the proposed approach.

Finally, the *buffer* is emptied to be prepared for the subsequent (next) window of upcoming instances ($wIter + 1$). A step-by-step pseudocode for the E-RAIDS base learner procedure is provided in Algorithm 1.

Algorithm 2 Ensemble of random feature subspaces

```
foreach wIter do
    foreach FS_i ∈ FS do
        subVote ← get subVote for wIter from subVoteList for FS_i
        eVote ← eVote + subVote for wIter in eVoteList
        subOutSet ← get subOutSet for wIter from subVoteList for FS_i
        append subOutSet to eOutSet for wIter in eVoteList
    end
    if eVote = vf_e then
        flag an alarm
    end
end
```

6.3.3.2 Ensemble of Random Feature Subspaces Voting

For the ensemble of p models $\{M_1, M_2, \ldots, M_n, M_p\}$ on feature subspaces $\{FS_1, FS_2, \ldots, FS_n, FS_p\}$ respectively, we define the following data repository and a vote factor:

- $eVoteList$: a list that stores the triples (1) a $wIter$, (2) an $eVote$ parameter that counts the number of feature subspaces that vote for an alarm, and (3) an outlier set $eOutSet$ that appends the $subOutSet$ from each FS_i votes for an alarm. It has the form $\langle wIter, subVote, subOutSet\rangle$.
- vf_e: a vote factor that confirms whether an alarm to be generated at $wIter$ by the whole ensemble. In other words, if a vf_e number of feature subspaces vote for an alarm, then an alarm is generated at $wIter$.

As illustrated in Fig. 6.2, the E-RAIDS aggregate component aggregates the results from the p feature subspaces, in order to confirm whether to generate an alarm or not at each window iteration $wIter$. For each feature subspace FS_i, if $subVote = 1$ for $wIter$, E-RAIDS adds $subVote$ to $eVote$ for $wIter$ in $eVoteList$. Furthermore, E-RAIDS gets $subOutSet$ for $wIter$ from $subVoteList$, and appends to $eOutSet$ for $wIter$ in $eVoteList$.

After getting the votes from all the feature subspaces in the ensemble, E-RAIDS checks if $eVote = vf_e$. If the condition is satisfied, then an alarm of a malicious insider threat is generated at $wIter$. The voting mechanism is technically controlled by the vote factor vf_e, such that if a vf_e number of feature subspaces vote for anomalous behaviour(s) at a window iteration $wIter$, then an alarm is generated. This accordingly handles the case of a conflict, where $\frac{p}{2}$ (50%) of the feature subspaces in the ensemble vote for anomalous behaviour(s) and the other $\frac{p}{2}$ vote for normal behaviour(s). The ensemble technically checks if $\frac{p}{2} = vf_e$, then an alarm is generated.

A step-by-step pseudocode for the E-RAIDS aggregate procedure is provided in Algorithm 2.

6.4 Experiments

We evaluated the effectiveness of the proposed approach on the CMU-CERT data sets using Windows Server 2016 on Microsoft Azure (RAM $140GB$, OS 64-$bits$, CPU Intel Xeon $E5 - 2673v3$) for data pre-processing and Microsoft Windows 7 Enterprise (RAM $12GB$, OS 64-$bits$, CPU Intel Core $i5$-4210U) for experiments. First, MATLAB $R2016b$ was used to pre-process the data set and generate community behaviour profiles per session slots of 4 h. Second, we implemented E-RAIDS-MCOD and E-RAIDS-AnyOut and carried out the experiments in Java environment (Eclipse Java Mars) using the open-source **MOA** package [33].

6.4.1 Description of the Data set

A significant impediment to researchers who work on the insider threat problem is the lack of real world data. The real data logs the activities executed by the users in an organisation. These data log files contain: private user profile information (e.g. name, email address, mobile number, home address, etc.); intellectual property (e.g. strategic or business plans, engineering or scientific information, source code, etc.); and confidential content (e.g. email content, file content, etc.) [7]. Organisations commonly refuse to give researchers access to real data to protect its users and assets.

In the current decade, there exists a great recent trend towards the utilisation of the CMU-CERT data set(s) for the insider threat detection systems [5, 12, 13, 30, 41, 43]. The CMU-CERT data sets are synthetic insider threat data sets generated by the CERT Division at Carnegie Mellon University [9, 16]. CMU-CERT data repository is the only one available for insider threat scenarios (5 scenarios) and has recently become the evaluation data repository for researchers addressing the insider threat problem [5, 17, 41].

In the preliminary version [22], we used $r5.1$ CMU-CERT data set, where each community consists of one malicious insider threat which map to one scenario.

For this chapter, we used $r5.2$ CMU-CERT data set which logs the behaviour of 2000 employees over 18 months. Unlike the previously released data sets, the communities in the $r5.2$ data set consist of multiple malicious insider threats which map to different scenarios. Among those employees, we extracted the data logs for employees belonging to the following three communities: Production line worker com-P, Salesman com-S, and IT admin com-I. The community com-P consists of 300 employees, 17 malicious insider threats (associated with 366 anomalous behaviours), where each threat maps to one of the scenarios $\{s1, s2, s4\}$. The community com-S consists of 298 employees and 22 malicious insider threats (associated with 515 anomalous behaviours), where each threat maps to one of the scenarios $\{s1, s2, s4\}$. The community com-I consists of 80 employees and 12 malicious insider threats (associated with 132 anomalous behaviours), where each threat maps to one of the scenarios $\{s2, s3\}$.

6.4.2 Experimental Tuning

In this work, we define two experiments based on the proposed E-RAIDS approach. The E-RAIDS-MCOD learns an ensemble of p MCOD base learners, such that each MCOD base learner is trained on a feature subspace of the p feature subspaces. Likewise, the E-RAIDS-AnyOut learns an ensemble of p AnyOut base learners.

The experiments are tuned for different values of parameters. Table 6.1 presents the values for the parameters tuned in each of MCOD and AnyOut, with a short description. Note that an extensive number of experiments were done to select the presented tuning values for the parameters. The values were selected based on E-RAIDS achieving the best performance in terms of the evaluation measures described below.

For MCOD, the parameter k has a default value $k = 50$ in the MOA package. In this chapter, we present $k = \{50, 60, 70\}$ to evaluate the E-RAIDS-MCOD for different number of neighbours. The parameter r, with a default value $r = 0.1$, is presented for a range for $r = \{0.3, 0.4, 0.5, 0.6, 0.7\}$ to check the influence of r.

For AnyOut, the parameter $|D_{train}|$ is presented for 500 instances, knowing that no threats are present at the beginning of the stream in these 500 instances. The parameters *confAgr* and *conf* did not show a significant influence on the utilised data sets, so both are assigned to their default values, in the MOA package, 2 and 4, respectively. τ has a minimum value 0 and a maximum value 1, so it is presented for $\tau = \{0.1, 0.4, 0.7\}$ to evaluate the influence of varying the outlier score threshold on the outliers detected. $oscAgr$ has a minimum value 1 and a maximum value 10, so it is presented for the $oscAgr = \{2, 4, 6, 8\}$.

In general, for E-RAIDS approach with either MCOD or AnyOut base learner, we present the vouch factor $vf_s = 2$, so it confirms an outlier as positive (anomalous) after it survives for 2 subsequent windows. We present the vote factor $vf_e = 1$, so if at least 1 feature subspace in the ensemble flags an alert, an alarm of a malicious insider threat is confirmed to be generated.

Likewise, for both E-RAIDS-MCOD and E-RAIDS-AnyOut, the window size is presented for $w = \{50, 100, 150, 200\}$. As previously described, an instance

Table 6.1 Tuned parameters

	Description
MCOD parameter	
$k = \{50, 60, 70\}$	Number of neighbours parameter
$r = \{0.3, 0.4, 0.5, 0.6, 0.7\}$	Distance parameter for outlier detection
AnyOut parameter	
$\|D_{train}\| = 500$	Training set size
$confAgr = 2$	Size of confidence aggregate
$conf = 4$	Initial confidence value
$\tau = \{0.1, 0.4, 0.7\}$	Outlier score threshold
$oscAgr = \{2, 4, 6, 8\}$	Size of outlier score aggregate

(feature vector) is a set of events executed during a pre-defined session slot. In this work, we define a session slot per 4 h. Let w represent the window size, which accumulates the acquired instances in a data stream. If $w = 50$ and $vf_s = 2$, then the instances in each window are processed after $4 \times 50 = 200\,\text{h} \simeq 8$ days. Based on the description of E-RAIDS, a threat (outlier) may be confirmed as an outlier at least over the window it belongs to (after 8 days) or over the next window (after $\simeq 8 \times 2 = 16$ days). If $w = 200$ and $vf_s = 2$, then the instances in each window are processed after $4 \times 200 = 800\,\text{h} \simeq 33$ days. A threat (outlier) may be confirmed as an outlier at least (after 33 days) or over the next window (after $\simeq 33 \times 2 = 66$ days). Note that the malicious insider threats simulated in the CMU-CERT data sets span over at least one month and up to 4 months. Hence, we hypothesise the following in Hypothesis 1.

Hypothesis 1 *The E-RAIDS approach is capable to detect the malicious insider threats in real-time (during the time span of the undergoing threat).*

6.4.3 Evaluation Measures

Many research work has been done to detect or mitigate malicious insider threats, but nevertheless has established standard measures to evaluate the proposed models [18]. The research practices show that the insider threat problem demands to measure the effectiveness of the models before being deployed, preferably in terms of true positives (TP) and false positives (FP) [19].

In the state of the art, a remarkable number of approaches were validated in terms of FP measure [4, 5, 14, 32, 37]. This sheds light on the importance of the FP measure to address the shortcoming of the high number of false alarms (FPs). Furthermore, some approaches were validated in terms of: TP measure [32]; F1 measure [4, 34]; AUC measure [8, 12, 17]; precision and recall [25, 30]; accuracy [25, 37]; and others.

The variety of the utilised evaluation measures in the state of the art reveals the critical need to formulate the insider threat problem and to define the measures that would best validate the effectiveness of the propose E-RAIDS approach. In the following, we give a formulation for the E-RAIDS approach and we define the evaluation measures utilised in this work.

As aforementioned, the ultimate aim of the E-RAIDS approach is to detect all the malicious insider threats over a data stream in real-time, while minimising the number of false alarms.

The challenge of the insider threat problem lies in the variety and complexity of the malicious insider threats in the data sets. Each malicious insider threat is devised of a set of anomalous behaviours. An anomalous behaviour is represented by an instance (feature vector) which describes a set of events (features) carried out by a malicious insider. Based on the challenge of the problem, we formulate the E-RAIDS approach with the aim to detect *any-behaviour-all-threat* (as defined in Sect. 6.1). This means that it is sufficient to detect any anomalous behaviour (instance) in all malicious insider threats. Hence, E-RAIDS approach is formulated as a *threat hunting* approach, where a threat is detected if (1) a vf_e number of feature subspaces confirm an undergoing threat (flag alarm) over a window and (2) the outliers, associated with the alarm flagged over the window, include at least a true positive (anomalous behaviour detected as outlier). Note that although the detection of one behaviour confirms the detection of the threat, we hypothesise the following in Hypothesis 2.

Hypothesis 2 *The E-RAIDS approach is capable of detecting more than one behaviour from the set of behaviours which belong to a malicious insider threat. We refer to as (more than one)-behaviour-all-threat detection in E-RAIDS.*

Furthermore, the property of the E-RAIDS approach of using windows over data streams introduces a refined version of the evaluation of false positives (FP). In this work, we use the FPAlarm to evaluate the false positives. So that if all the outliers associated with the alarm generated over a window are actually normal, then the alarm is considered a false alarm (i.e. FPAlarm is incremented 1). A formal definition for FPAlarm is given in the following.

We define the measures used to evaluate the performance of E-RAIDS, which include a refined version of the default (known) evaluation measures, taking our ultimate aim into account.

- **P**$_T$: *Threats* number of malicious insider threats associated with anomalous instances. In other words, P_T is the number of malicious insiders attributed to the anomalous behaviours;
- **TP**$_T$: *True Positives* a refined version of default TP to evaluate the number of threats detected by the framework among all the P_T malicious insider threats. TP_T is incremented if at least one anomalous instance (behaviour attributed to the threat) is associated as an outlier to a flagged alarm;
- **FPAlarm**: *False Positive Alarm* a refined version of default FP to evaluate the number of false alarms generated. An alarm is declared false if all the outliers associated with the alarm are actually normal instances. This means that none of the instances contributed to generating the alarm, therefore, false alarm;
- **FN**$_T$: *False Negatives* a refined version of default FN to evaluate the number of insider threats not detected; and
- **F1** measure: defined based on the values of the above defined measures.

The evaluation for E-RAIDS does not employ only the defined evaluation measures, however, it is required to prove the previously stated hypothesises to be true.

6.4.4 *Results and Discussion*

In the preliminary version [22], RandSubOut was evaluated in terms of TP_T detected threats and FPAlarm.

In this work, the results are presented and discussed for E-RAIDS-MCOD and E-RAIDS-AnyOut as follows: in terms of (1) the pre-defined evaluation measures; (2) voting feature subspaces; (3) real-time anomaly detection; and (4) the detection of (more than One)-behaviour-all-threat.

6.4.4.1 MCOD vs AnyOut Base Learner for E-RAIDS in Terms of Evaluation Measures

In the following, we analyse the performance of E-RAIDS with MCOD base learner vs AnyOut base learner in terms of the pre-defined evaluation measures: TP_T out of P_T; FPAlarm; and F1 measure.

Tables 6.2 and 6.3 present the maximum TP_T and the minimum FPAlarm attained E-RAIDS-MCOD and E-RAIDS-AnyOut over the communities. The results are reported in terms of the parameter values in the given sequence k, r, w for E-RAIDS-MCOD and $\tau, oscAgr, w$ for E-RAIDS-AnyOut, respectively.

E-RAIDS-MCOD

Figure 6.3 presents the variation of $F1$ measure as a function of window size w for E-RAIDS with MCOD base learner over the communities. The results are reported with respect to k and r parameter values.

A preliminary analysis of the F1 measure shows no evident pattern in terms of any of the parameters k, r, or w. Over the community com-P, E-RAIDS-MCOD

Table 6.2 Maximum TP_T of detected insider threats over communities

Community	E-RAIDS-MCOD	Parameters k, r, w
com-P	**16**	50,0.3,100 60,0.4,50
		60,0.6,100 60,0.7,150
		70,0.4,50
com-S	**21**	70,0.4,150
com-I	10	50,0.4,50 70,0.3,100
	E-RAIDS-AnyOut	Parameters $\tau, oscAgr, w$
com-P	**16**	0.1,2,100–200 {0.3, 0.7},2,50–200
com-S	20	0.1,2,50–100 0.3,2,50–100
		0.7,2,150
com-I	**12**	0.1,2,50–200 0.3,2,50–100
		0.7,2,{50, 200}

The bold values represent the maximum TPT achieved by either E-RAIDS-MCOD or E-RAIDS-AnyOut over each community

Table 6.3 Minimum FPAlarm over communities

Community	E-RAIDS-MCOD	Parameters k, r, w
com-P	**0**	50,0.4,200 60,{0.3, 0.5, 0.6},200
com-S	**0**	50,0.4,200 50,0.7,100
		60,0.3,200 60,0.5–0.7,100
		70,0.6–0.7,150
com-I	**0**	50,0.3,200 60,0.5,200
		70,0.5–0.7,200
	E-RAIDS-AnyOut	**Parameters τ, $oscAgr$, w**
com-P	2	$\forall\tau$,$\forall oscAgr$,200
com-S	2	$\forall\tau$,$\forall oscAgr$,150–200
com-I	1	0.1,{2, 4},200 0.3,2–6,200
		0.7,2–4,200

The bold values represent the minimum FPAlarm by either E-RAIDS-MCOD or E-RAIDS-AnyOut over each community

achieves the maximum F1 = 0.9411; 60, 0.7, 150. It detects the maximum $TP_T = 16$ out of $P_T = 17$, thus missing one malicious insider threat. However, it flags only a false positive alarm (FPAlarm = 1). Furthermore, Table 6.3 shows that E-RAIDS-MCOD reports the minimum FPAlarm = 0 at $w = 200$ for different values of k and r.

Over the community com-S, E-RAIDS-MCOD achieves the maximum F1= 0.9523; 70, {0.6, 0.7}, 150. It detects a $TP_T = 20$ out of $P_T = 22$, while flagging no false positive alarms (FPAlarm= 0). The maximum $TP_T = 21$ is attained at 70, 0.4, 150, however, flagging FPAlarm= 2.

Over the community com-I, E-RAIDS-MCOD achieves the maximum F1= 0.6451; 70, 0.3, 100. It detects a $TP_T = 10$ out of $P_T = 12$, thus missing two malicious insider threats, while flagging FPAlarm= 9. Nevertheless, Table 6.3 shows that E-RAIDS-MCOD reports the minimum FPAlarm= 0 at $w = 200$ for different values of k and r.

We can deduce that the window size $w = 150, 200$ gives the best performance for E-RAIDS-MCOD in terms of the evaluation measures.

E-RAIDS-AnyOut

Figure 6.4 presents the variation of F1 measure as a function of window size w for E-RAIDS with AnyOut base learner over the communities. The results are reported with respect to τ and $oscAgr$ parameter values.

It is evident that there exists a positive correlation between F1 measure and the parameter $oscAgr$ for E-RAIDS-AnyOut. The variation of F1 measure at $oscAgr = 2$ is the highest with respect to all the window sizes $w =$ 50–200 over both communities. Moreover, Fig. 6.5a reveals a positive correlation between F1 measure and the parameter w. The value of F1 measure increases as the window size w increases.

Over the community com-P, E-RAIDS-AnyOut achieves the maximum F1= 0.9142; $\forall\tau$, 2, 200. It detects the maximum $TP_T = 16$ out of $P_T = 17$, while

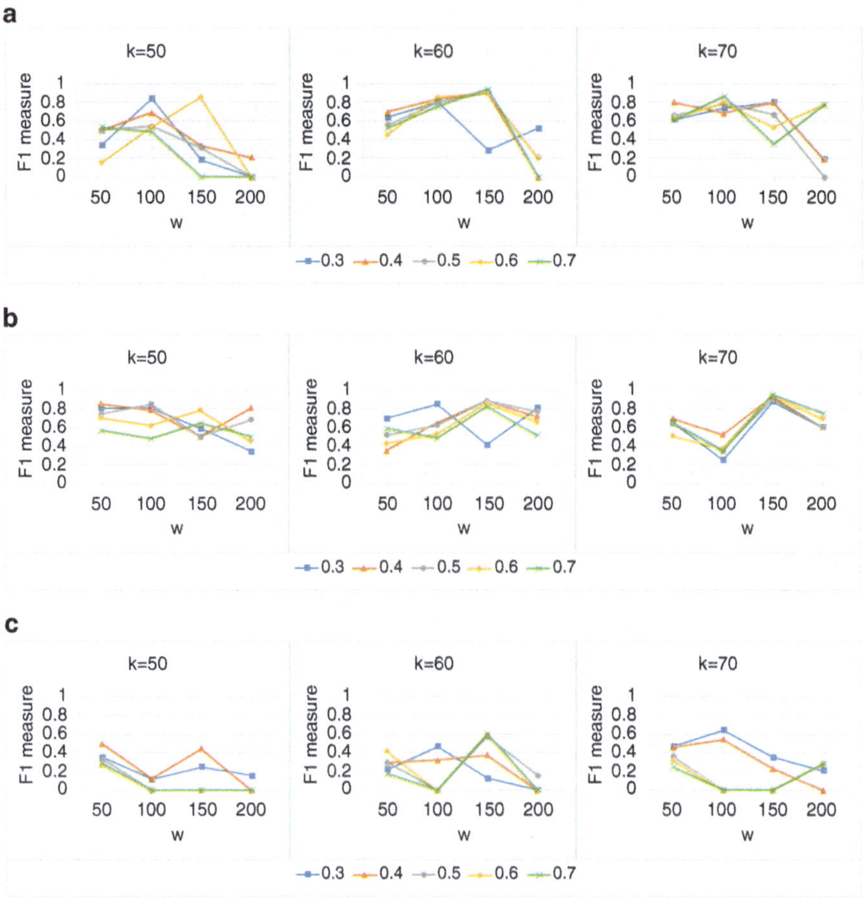

Fig. 6.3 The variation of F1 measure as a function of window size w for E-RAIDS with MCOD base learner over the communities. The legend represents the r parameter values. (**a**) com-P. (**b**) com-S. (**c**) com-I

reducing the false positive alarms to FPAlarm= 2. Table 6.3 shows that E-RAIDS-AnyOut reports the minimum FPAlarm= 2 $\forall\tau, \forall oscAgr$ at $w = 200$. Thus, the higher the window size, the lower the FPAlarm. Table 6.2 shows that E-RAIDS-AnyOut reports the maximum $TP_{\mathcal{T}} = 16$ at $oscAgr = 2$ in general terms $\forall\tau, \forall w$. Thus, the lower the $oscAgr$, the higher the $TP_{\mathcal{T}}$ detected.

Over the community com-S, E-RAIDS-AnyOut achieves the maximum F1= 0.9090; $0.7, 2, 150$. It detects a $TP_{\mathcal{T}} = 20$ out of $P_{\mathcal{T}} = 22$, while flagging two false positive alarms (FPAlarm= 2).

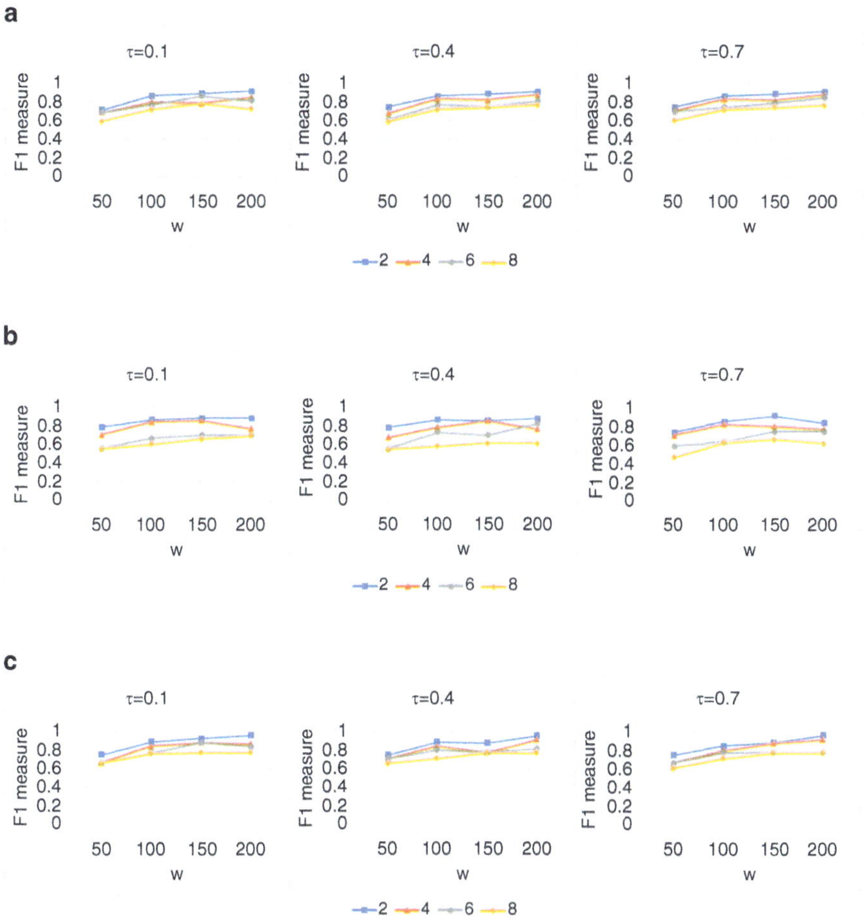

Fig. 6.4 The variation of F1 measure as a function of window size w for E-RAIDS with AnyOut base learner over the communities. The legend represents the $oscAgr$ parameter values. (**a**) com-P. (**b**) com-S. (**c**) com-I

Over the community com-I, E-RAIDS-AnyOut achieves the maximum F1= 0.9600; $\forall\tau, 2, 200$. It detects a $\text{TP}_T = 12$ out of $\text{P}_T = 12$, while flagging one false positive alarm (FPAlarm= 1).

In terms of the evaluation measures, E-RAIDS-MCOD outperforms E-RAIDS-AnyOut over the communities, where E-RAIDS-MCOD achieves a higher F1 measure over com-P and com-S, a higher TP_T over com-S, and a lower FPAlarm= 0 over all communities.

6.4.4.2 MCOD vs AnyOut for E-RAIDS in Terms of Voting Feature Subspaces

In the following, we address the merit of using feature subspaces in the E-RAIDS approach. We compare the number of feature subspaces in the ensemble that voted for a malicious insider threat in each of E-RAIDS-MCOD and E-RAIDS-AnyOut. The rationale behind using a random feature subspace to train each model of the p models in the ensemble is to train the base learner (MCOD or AnyOut) on a subset of the features and to detect local outliers which might not be detected over the whole feature space.

Figure 6.5 illustrates the number of votes which contributed in flagging an alarm of a malicious threat with respect to the number of instances processed (given the window size w) over the communities. The number of votes actually corresponds to the number of feature subspaces in the ensemble which generated an alert. We selected E-RAIDS-MCOD and E-RAIDS-AnyOut with their parameters which showed the best performance in terms of the evaluation measures. For move

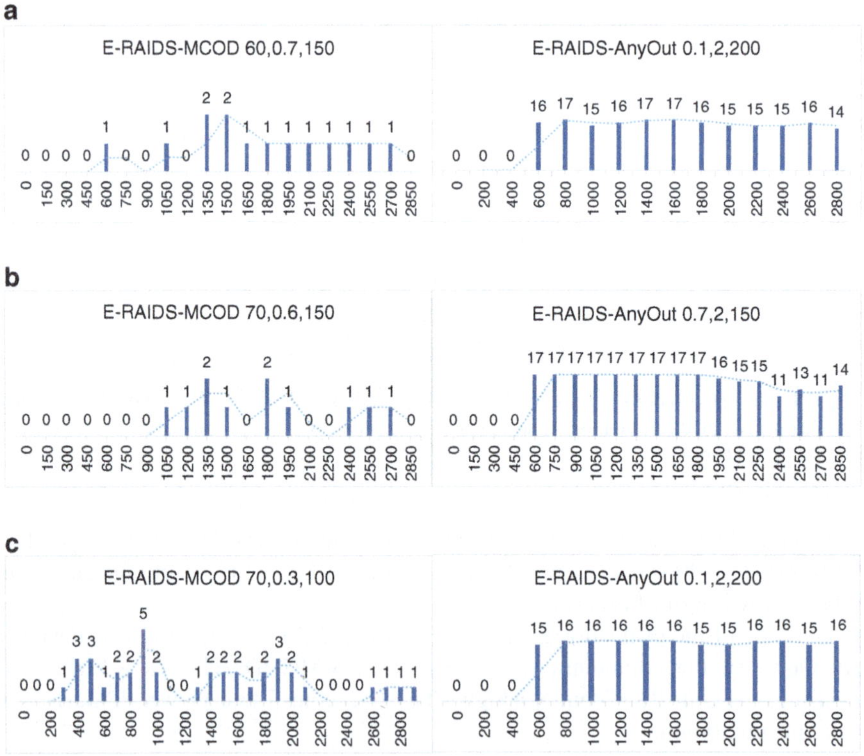

Fig. 6.5 The number of votes which contributed in flagging an alarm of a malicious threat with respect to the number of instances processed over the communities. (**a**) com-P. (**b**) com-S. (**c**) com-I

com-P down to be clear Fig. 6.5a shows the E-RAIDS-MCOD at 60, 0.7, 150 and E-RAIDS-AnyOut at 0.1, 2, 200. For com-S, Fig. 6.5b shows the E-RAIDS-MCOD at 70, 0.6, 150 and E-RAIDS-AnyOut at 0.7, 2, 150. For com-I, Fig. 6.5c shows the E-RAIDS-MCOD at 70, 0.3, 100 and E-RAIDS-AnyOut at 0.1, 2, 200. Given that the $|D_{train}| = 500$, this justifies why the votes for E-RAIDS-AnyOut start after 500 train instances has been processed over both communities.

Given the window size w, a window iteration $wIter$ starts at the number of already processed instances $procInst$ and ends at $procInst+w$. For example, given $w = 100$, the first window iteration $wIter = 0$ starts at $procInst = 0$ and ends at $procInst + w = 100$; $wIter = 1$ starts at 100 and ends at 200; etc.

A preliminary analysis of the results shows that E-RAIDS-AnyOut flags alarms continuously $\forall wIter$ after $procInst = 500$. However, E-RAIDS-MCOD shows distinct alarms flagged, with no alarms at certain windows iterations. We recall that E-RAIDS-MCOD outperforms E-RAIDS-AnyOut in terms of FPAlarm measure. The continuous alarms flagged $\forall wIter$ in E-RAIDS-AnyOut explains the higher FPAlarm, as well as it reveals the uncertainty of E-RAIDS-AnyOut compared to E-RAIDS-MCOD.

Knowing that the number of feature subspaces utilised in the ensemble is $p = 17$, the number of votes $\forall wIter$ in E-RAIDS-AnyOut has a minimum $votes = 11$ and a maximum $votes = 17$, which reveals a level of uncertainty. The case where 17 feature subspaces vote for an alarm indicates that all (17) models of the ensemble detect *at least one* outlier (positive) associated with a malicious insider threat. On the other hand, the number of votes in E-RAIDS-MCOD has a maximum $votes = 2$. This means that only 1 or 2 feature subspaces vote for an alarm. We recall the complexity of the malicious insider threat scenarios in the CMU-CERT data sets. The anomalous instances usually exist in (sparse or dense) regions of normal instances, and rarely as *global outliers* with respect to the whole feature space. To address this, the E-RAIDS approach aims to detect *local outliers* which may be found over ANY (not ALL) of feature subspaces. Having all the feature subspaces in E-RAIDS-AnyOut voting for a threat, compared to a couple (1 or 2) of feature subspaces in E-RAIDS-MCOD, reinforces the uncertainty of E-RAIDS-AnyOut. The reason behind the uncertain performance of AnyOut in the E-RAIDS approach may be due to that the outlier score of an instance X^t is computed upon the arrival of a new instance X^{t+1}. Thus, the processing of the instance X^t is interrupted at a certain level of the ClusTree, and the outlier score is computed with a lower level of confidence (i.e. uncertain).

6.4.4.3 Real-Time Anomaly Detection in E-RAIDS

To prove the aforementioned Hypothesis 1 to be true, it is required to check if the E-RAIDS detects the malicious insider threats in real-time (where real-time means that the alarm is flagged during the time span of the undergoing threat). Based on the previous conclusion regarding the uncertainty of E-RAIDS-AnyOut, and the

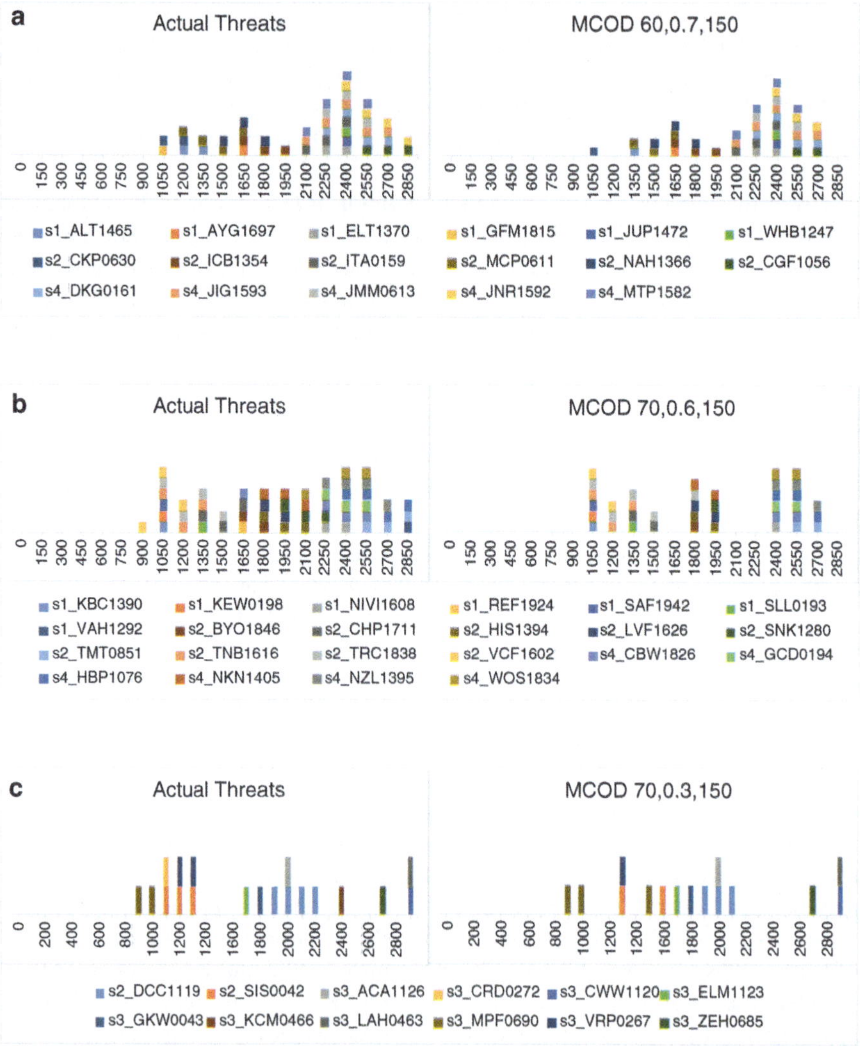

Fig. 6.6 The actual malicious insider threats vs the threats detected in E-RAIDS-MCOD with respect to the number of instances processed over the communities. (**a**) com-P. (**b**) com-S. (**c**) com-I

superiority of E-RAIDS-MCOD in terms of (1) the evaluation measures and (2) the voting feature subspaces, we select E-RAIDS-MCOD to verify Hypothesis 1.

Figure 6.6 illustrates the actual malicious insider threats vs the threats detected in E-RAIDS-MCOD with respect to the number of instances processed over the communities. The malicious insider threats are displayed in the legend over each community using the following label scenRef_insiderID (e.g. s1_ALT1465) such that scenRef (e.g. s1, s2, or s4) represents the reference number for the scenario followed in the malicious insider threat; and insiderID (e.g. ALT1465, AYG1697) represents the user ID of the insider attributed to the threat. Hence, each colour in the legend refers to a malicious insider threat.

We observe in Fig. 6.6 that a malicious insider threat is detected either (obs1) at the current window iteration $wIter$ where it is actually simulated, or (obs2) at the subsequent (next) $wIter$. Hence, Hypothesis 1 is verified.

Based on the description of the E-RAIDS approach, a feature subspace votes for a threat at $wIter$ if at least an outlier survived for a vf_s number of subsequent windows, and consequently a specific threat is detected at $wIter$ if (cond1) a vf_e number of subspaces vote (an alarm is flagged), and (cond2) at least outlier (positive) associated with the alarm belongs to the threat. However, the outliers associated with the alarm (as mentioned in (cond2)) consist of *persistent* outliers (which survived from $wIter - 1$) and *potential* outliers (at the current $wIter$). A potential outlier, if satisfies (cond2), allows real-time detection at the current $wIter$ (obs1). A persistent outlier, if satisfies condition (cond2), allows real-time detection as observed at the subsequent $wIter$ (obs2).

6.4.4.4 (More Than One)-Behaviour-All-Threat Detection in E-RAIDS

The final analysis addresses Hypothesis 2. As defined in Sect. 6.1, the idea of *threat hunting* aims to detect any-behaviour-all-threat, however, Fig. 6.6 shows the capability of E-RAIDS-MCOD to detect more than one behaviour (not only one) from the set of behaviours which belong to a malicious insider threat. It manifests as multiple alarms (colour spikes) generated for a specific threat over a number of windows. Hence, Hypothesis 2 is verified.

Analytically, this underlies in having multiple outliers, associated with the alarm(s) flagged over window(s), which are actually true positives belonging to a specific malicious insider threat.

6.5 Conclusion and Future Work

This chapter addresses the shortcoming of high number of false alarms in the existing insider threat detection mechanisms. The continuous flagging of false alarms deceives the administrator(s) about suspicious behaviour of many users. This consumes a valuable time from their schedule, while investigating the suspected users.

We present a streaming anomaly detection approach, namely Ensemble of Random subspace Anomaly detectors In Data Streams (E-RAIDS), for insider threat detection. The ultimate aim of E-RAIDS is to detect any-behaviour-all-threat (threat hunting as defined in Sect. 6.1), while reducing the number of false alarms.

E-RAIDS is built on top of established continuous outlier detection techniques (MCOD or AnyOut). These techniques use *data stream clustering* to optimise the detection of outliers (which may refer to malicious insider threats). E-RAIDS is an ensemble of p outlier detectors (p MCOD base learners or p AnyOut base learners), where each model of the p models learns on a random feature subspace. The merit of using feature subspaces is to detect local outliers which might not be detected over the whole feature space. These outliers may refer to anomalous behaviour(s) which belong to a malicious insider threat. E-RAIDS presents also an aggregate component to combine the votes from the feature subspaces, and take a decision whether to flag an alarm or not.

We define two experiments: E-RAIDS-MCOD with MCOD base learner, and E-RAIDS-AnyOut with AnyOut base learner. The experiments are carried out on CMU-CERT data sets which include simulated malicious insider threat scenarios. We compare the performance of E-RAIDS using each of MCOD and AnyOut in terms of: (1) the evaluation measures: F1 measure, TP_t of threats detected, and FPAlarm flagged; (2) the effectiveness of the concept of voting feature subspaces; (3) the capability of E-RAIDS to detect insider threats in real-time (Hypothesis 1); and (4) the capability of E-RAIDS to detect more than one behaviour belonging to an insider threat (Hypothesis 2) despite our formulation to the insider threat approach (threat hunting).

The results show that E-RAIDS-MCOD outperforms E-RAIDS-AnyOut, where the latter shows a low level of certainty in the detection of outliers. E-RAIDS-MCOD reports a higher F1 measure $=$ 0.9411 and 0.9523 over com-P and com-S, a lower FPAlarm $=$ 0 over all communities, and misses only one threat $TP_T = 16$ and 21 over com-P and com-S. It is worth to also mention that the window size $w = 150, 200$ gives the best performance for E-RAIDS-MCOD compared to the tuned values. E-RAIDS verifies the hypothesised capabilities in terms of the detection of more than one behaviour per threat in real-time.

References

1. C.C. Aggarwal, S.Y. Philip, J. Han, J. Wang, A framework for clustering evolving data streams, in *Proceedings 2003 VLDB Conference* (Elsevier, Burlington, 2003), pp. 81–92
2. F. Angiulli, F. Fassetti, Distance-based outlier queries in data streams: the novel task and algorithms. Data Min. Knowl. Disc. **20**(2), 290–324 (2010)
3. I. Assent, P. Kranen, C. Baldauf, T. Seidl, Anyout: anytime outlier detection on streaming data, in *International Conference on Database Systems for Advanced Applications* (Springer, Berlin, 2012), pp. 228–242
4. A. Azaria, A. Richardson, S. Kraus, V. Subrahmanian, Behavioral analysis of insider threat: a survey and bootstrapped prediction in imbalanced data. IEEE Trans. Comput. Soc. Syst. **1**(2), 135–155 (2014)
5. B. Böse, B. Avasarala, S. Tirthapura, Y.Y. Chung, D. Steiner, Detecting insider threats using radish: a system for real-time anomaly detection in heterogeneous data streams. IEEE Syst. J. **11**(2), 471–482 (2017)
6. M.M. Breunig, H.P. Kriegel, R.T. Ng, J. Sander, LOF: identifying density-based local outliers, in *ACM Sigmod Record*, vol. 29 (ACM, New York, 2000), pp. 93–104

7. D.M. Cappelli, A.P. Moore, R.F. Trzeciak, *The CERT Guide to Insider Threats: How to Prevent, Detect, and Respond to Information Technology Crimes (Theft, Sabotage, Fraud)* (Addison-Wesley, Upper Saddle River, 2012)
8. Y. Chen, S. Nyemba, B. Malin, Detecting anomalous insiders in collaborative information systems. IEEE Trans. Dependable Secure Comput. **9**(3), 332–344 (2012)
9. CMU CERT Team, CMU cert synthetic insider threat data set. https://resources.sei.cmu.edu/library/asset-view.cfm?assetid=508099. Accessed 12 Apr 2018
10. CMU Software Engineering Institute, 2011 cybersecurity watch survey. https://www.sei.cmu.edu/news/article.cfm?assetid=52441. Accessed 14 Feb 2018
11. M. Ester, H.P. Kriegel, J. Sander, X. Xu et al., A density-based algorithm for discovering clusters in large spatial databases with noise, in *Kdd*, vol. 96 (1996), pp. 226–231
12. A. Gamachchi, S. Boztas, Insider threat detection through attributed graph clustering, in *2017 IEEE Trustcom/BigDataSE/ICESS* (IEEE, Piscataway, 2017), pp. 112–119
13. A. Gamachchi, L. Sun, S. Boztas, Graph based framework for malicious insider threat detection, in *Proceedings of the 50th Hawaii International Conference on System Sciences*, Hawaii, 4–7 January 2017, pp. 2638–2647
14. C. Gates, N. Li, Z. Xu, S.N. Chari, I. Molloy, Y. Park, Detecting insider information theft using features from file access logs, in *European Symposium on Research in Computer Security* (Springer, Switzerland, 2014), pp. 383–400
15. D. Georgiadis, M. Kontaki, A. Gounaris, A.N. Papadopoulos, K. Tsichlas, Y. Manolopoulos, Continuous outlier detection in data streams: an extensible framework and state-of-the-art algorithms, in *Proceedings of the 2013 ACM SIGMOD International Conference on Management of Data* (ACM, New York, 2013), pp. 1061–1064
16. J. Glasser, B. Lindauer, Bridging the gap: a pragmatic approach to generating insider threat data, in *2013 IEEE Security and Privacy Workshops (SPW)* (IEEE, Piscataway, 2013), pp. 98–104
17. H. Goldberg, W. Young, M. Reardon, B. Phillips et al., Insider threat detection in prodigal, in *Proceedings of the 50th Hawaii International Conference on System Sciences* (2017)
18. F.L. Greitzer, T.A. Ferryman, Methods and metrics for evaluating analytic insider threat tools, in *2013 IEEE Security and Privacy Workshops (SPW)* (IEEE, Piscataway, 2013), pp. 90–97
19. M.D. Guido, M.W. Brooks, Insider threat program best practices, in *2013 46th Hawaii International Conference on System Sciences (HICSS)* (IEEE, Piscataway, 2013), pp. 1831–1839
20. A. Guttman, *R-Trees: A Dynamic Index Structure for Spatial Searching*, vol. 14 (ACM, New York, 1984)
21. M. Hahsler, M. Bolanos, J. Forrest et al., Introduction to stream: an extensible framework for data stream clustering research with R. J. Stat. Softw. **76**(14), 1–50 (2017)
22. D. Haidar, M.M. Gaber, Outlier detection in random subspaces over data streams: an approach for insider threat detection. Expert Update **17**(1), 1–16 (2017)
23. D.M. Hawkins, *Identification of Outliers*, vol. 11 (Springer, Dordrecht, 1980)
24. Z. He, X. Xu, S. Deng, Discovering cluster-based local outliers. Pattern Recogn. Lett. **24**(9–10), 1641–1650 (2003)
25. M. Kandias, V. Stavrou, N. Bozovic, D. Gritzalis, Proactive insider threat detection through social media: the youtube case, in *Proceedings of the 12th ACM Workshop on Privacy in the Electronic Society* (ACM, New York, 2013), pp. 261–266
26. M. Khalilian, N. Mustapha, Data stream clustering: challenges and issues (2010, Preprint). arXiv: 1006.5261
27. E.M. Knox, R.T. Ng, Algorithms for mining distance based outliers in large datasets, in *Proceedings of the International Conference on Very Large Data Bases*. Citeseer (1998), pp. 392–403
28. M. Kontaki, A. Gounaris, A.N. Papadopoulos, K. Tsichlas, Y. Manolopoulos, Continuous monitoring of distance-based outliers over data streams, in *2011 IEEE 27th International Conference on Data Engineering* (IEEE, Piscataway, 2011), pp. 135–146

29. P. Kranen, I. Assent, C. Baldauf, T. Seidl, Self-adaptive anytime stream clustering, in *Ninth IEEE International Conference on Data Mining, ICDM'09* (IEEE, Piscataway, 2009), pp. 249–258
30. P.A. Legg, O. Buckley, M. Goldsmith, S. Creese, Automated insider threat detection system using user and role-based profile assessment. IEEE Syst. J. **11**, 503–512 (2015)
31. P.A. Legg, O. Buckley, M. Goldsmith, S. Creese, Caught in the act of an insider attack: detection and assessment of insider threat, in *2015 IEEE International Symposium on Technologies for Homeland Security (HST)* (IEEE, Piscataway, 2015), pp. 1–6
32. M. Mayhew, M. Atighetchi, A. Adler, R. Greenstadt, Use of machine learning in big data analytics for insider threat detection, in *MILCOM 2015–2015 IEEE Military Communications Conference* (IEEE, Piscataway, 2015), pp. 915–922
33. MOA Team, T.U.o.W., Massive online analysis open source framework. https://moa.cms.waikato.ac.nz/. Accessed 14 Feb 2018
34. P. Moriano, J. Pendleton, S. Rich, L.J. Camp, Insider threat event detection in user-system interactions, in *Proceedings of the 2017 International Workshop on Managing Insider Security Threats* (2017)
35. J.R. Nurse, P.A. Legg, O. Buckley, I. Agrafiotis, G. Wright, M. Whitty, D. Upton, M. Goldsmith, S. Creese, A critical reflection on the threat from human insiders—its nature, industry perceptions, and detection approaches, in *International Conference on Human Aspects of Information Security, Privacy, and Trust* (Springer, Cham, 2014), pp. 270–281
36. P. Parveen, Z.R. Weger, B. Thuraisingham, K. Hamlen, L. Khan, Supervised learning for insider threat detection using stream mining, in *2011 IEEE 23rd International Conference on Tools with Artificial Intelligence* (IEEE, Piscataway, 2011), pp. 1032–1039
37. P. Parveen, N. Mcdaniel, Z. Weger, J. Evans, B. Thuraisingham, K. Hamlen, L. Khan, Evolving insider threat detection stream mining perspective. Int. J. Artif. Intell. Tools **22**(05) (2013). https://doi.org/10.1142/S0218213013600130
38. Reuters, Harold marin theft of ip. http://www.reuters.com/article/us-usa-cybersecurity-nsa-contractor-idUSKBN15N2N4. Accessed 14 Feb 2018
39. T. Seidl, I. Assent, P. Kranen, R. Krieger, J. Herrmann, Indexing density models for incremental learning and anytime classification on data streams, in *Proceedings of the 12th International Conference on Extending Database Technology: Advances in Database Technology* (ACM, New York, 2009), pp. 311–322
40. J.A. Silva, E.R. Faria, R.C. Barros, E.R. Hruschka, A.C. De Carvalho, J. Gama, Data stream clustering: a survey. ACM Comput. Surv. (CSUR) **46**(1), 13 (2013)
41. A. Tuor, S. Kaplan, B. Hutchinson, N. Nichols, S. Robinson, Deep learning for unsupervised insider threat detection in structured cybersecurity data streams (2017) arXiv preprint arXiv:1710.00811
42. J. Verble, The NSA and Edward Snowden: surveillance in the 21st century. ACM SIGCAS Comput. Soc. **44**(3), 14–20 (2014)
43. S. Walton, E. Maguire, M. Chen, Multiple queries with conditional attributes (qcats) for anomaly detection and visualization, in *Proceedings of the Eleventh Workshop on Visualization for Cyber Security* (ACM, New York, 2014), pp. 17–24
44. D. Yang, E.A. Rundensteiner, M.O. Ward, Neighbor-based pattern detection for windows over streaming data, in *Proceedings of the 12th International Conference on Extending Database Technology: Advances in Database Technology* (ACM, New York, 2009), pp. 529–540
45. A. Zargar, A. Nowroozi, R. Jalili, Xaba: a zero-knowledge anomaly-based behavioral analysis method to detect insider threats, in *2016 13th International Iranian Society of Cryptology Conference on Information Security and Cryptology (ISCISC)* (IEEE, Piscataway, 2016), pp. 26–31
46. T. Zhang, R. Ramakrishnan, M. Livny, Birch: an efficient data clustering method for very large databases, in *ACM Sigmod Record*, vol. 25 (ACM, New York, 1996), pp. 103–114

Chapter 7
Effective Tensor-Based Data Clustering Through Sub-Tensor Impact Graphs

K. Selçuk Candan, Shengyu Huang, Xinsheng Li, and Maria Luisa Sapino

7.1 Introduction

Tensors are multi-dimensional arrays and are commonly used for representing multi-dimensional data, such as sensor streams and social networks [9, 17]. Thanks to the widespread availability of multi-dimensional data, tensor decomposition operations (such as CP [10] and Tucker [31]) are increasingly being used to implement various data analysis tasks, from anomaly detection [17], correlation analysis [26] to pattern discovery [13] and clustering [22, 28, 32].

A critical challenge for tensor-based analysis is its computational complexity and decomposition can be a bottleneck in some applications [14, 21, 30]. Phan and Cichocki [23] proposed a methodology to partition the tensor into smaller sub-tensors to deal with this issue: (a) partition the given tensor into blocks (or sub-tensors), (b) decompose each block independently, and then (c) iteratively combine these sub-tensor decompositions into a final decomposition for the input tensor. This process leads to two key observations:

- Observation #1: Our key observation in this chapter is that *Step (c), which iteratively updates and stitches the sub-tensor decompositions obtained in Steps (a) and (b), is where the various decompositions interact with each other and*

K. S. Candan (✉) · S. Huang · X. Li
Arizona State University, Tempe, AZ, USA
e-mail: candan@asu.edu; shengyu.huang@asu.edu; lxinshen@asu.edu

M. L. Sapino
University of Torino, Torino, Italy
e-mail: marialuisa.sapino@unito.it

O. Nasraoui, C.-E. Ben N'Cir (eds.), *Clustering Methods for Big Data Analytics*, Unsupervised and Semi-Supervised Learning,
https://doi.org/10.1007/978-3-319-97864-2_7

where any inaccuracies in individual sub-tensor decompositions can propagate (through the update rules introduced in Sect. 7.2.4) to the decomposition of the complete tensor.

- Observation #2: We further observe that *if we can quantify and capture how these sub-tensors interact and inaccuracies propagate, we can use this information to better allocate resources to tackle the accuracy–efficiency trade-off inherent in the decomposition process.*

Based on these two observations, in this chapter, we introduce the notion of *sub-tensor impact graphs (SIGs)* (Sect. 7.3), which capture and represent how the decompositions of these sub-partitions impact each other and the overall tensor decomposition accuracy and present several complementary algorithms that leverage this novel concept to address various key challenges in tensor decomposition.

7.1.1 Contributions of This Chapter: Sub-Tensor Impact Graphs

While block-based tensor decomposition techniques [15, 23] provide potential opportunities to boost the accuracy/efficiency trade-off, this solution leaves several open questions, including (a) how to partition the tensor and (b) how to most effectively combine results from these partitions. In this chapter, we introduce the notion of *sub-tensor impact graphs (SIGs)*, which quantify how the decompositions of these sub-partitions impact each other and the overall tensor decomposition accuracy and present four complementary algorithms that leverage this novel concept to address various key challenges in tensor decomposition, including *personalization*, *noise*, and *dynamic data*.

7.1.1.1 Challenge #1: Decomposition in the Presence of Dynamic Data

Firstly, we rely on sub-tensor impact graphs (SIGs) to tackle performance challenges that dynamic data pose in tensor analytics: incremental tensor decomposition. Re-computation of the whole tensor decomposition with each update will cause high computational costs and incur large memory overheads. Especially for applications where data evolves over time and the tensor-based analysis results need to be continuously maintained. In Sect. 7.4, we present a two-phase block-incremental CP-based tensor decomposition technique (BICP), which relies on sub-tensor impact graphs to prune unnecessary computation in the presence of incremental updates on the data [11].

7.1.1.2 Challenge #2: Dealing with Noisy Data

Next, in Sect. 7.5, we present a *Noise Adaptive Tensor Decomposition* (`nTD`) method that leverages sub-tensor impact graphs to tackle deal with noisy data. `nTD` partitions the tensor into multiple sub-tensors and then decomposes each sub-tensor probabilistically through Bayesian factorization—the resulting decompositions are then recombined through an iterative refinement process to obtain the decomposition for the whole tensor. `nTD` leverages a resource allocation strategy that accounts for the impact of the noise density of one sub-tensor on the decomposition accuracies of the other sub-tensors, based on the underlying sub-tensor impact graph [19].

7.1.1.3 Challenge #3: Personalization of the Decomposition Process

Finally, we introduce a novel personalized tensor decomposition (PTD) mechanism for accounting for the user's focus and interests during tensor decomposition (Sect. 7.6). We present alternative ways to account for the impact of the accuracy of one region of the tensor to the accuracies of the other regions of the tensor, each based on a different assumption about how the impact of inaccuracies propagates along the tensor. Given a model of impact, PTD (a) first partitions the input tensor in a way that reflects user's interest, (b) constructs a sub-tensor impact graph reflecting the tensor content and its partitions, and then (c) analyzes this sub-tensor impact graph (in the light of the user's interest) to identify initial decomposition ranks for the sub-tensors in a way that will boost the final decomposition accuracies for partitions of interest [18].

7.2 Background

7.2.1 Tensors

A tensor is a multi-dimensional array. More formally, an N-way or Nth-order tensor is an element of the tensor product of N vector spaces, each of which has its own coordinate system. A third-order tensor has three indices. A first-order tensor is a vector, a second-order tensor is a matrix, and tensors of order three or higher are called higher-order tensors. As in the case of matrices, the dimensions of the tensor array are referred to as its modes. For example, the tensor, $\mathcal{X} \in \mathbb{R}^{I \times J \times K}$, shown in Fig. 7.1, is of third-order and has three modes: I columns (mode 1), J rows (mode 2), and K tubes (mode 3). Fibers are the higher-order analogue of matrix rows and columns. A fiber is defined by fixing every index but one. A matrix column is a mode-1 fiber and a matrix row is a mode-2 fiber. Slices are two-dimensional sections of a tensor, defined by fixing all but two indices [16].

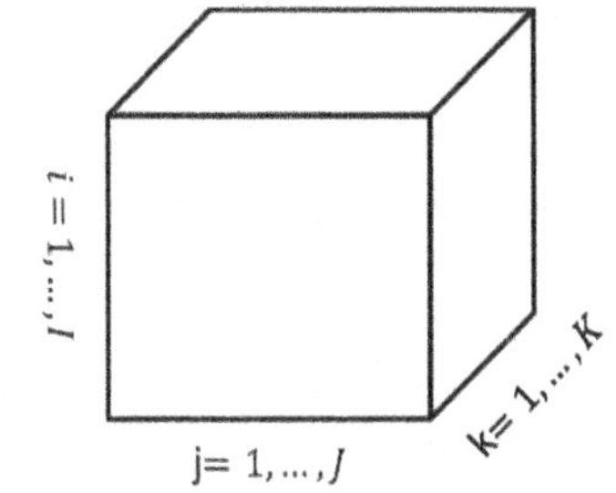

Fig. 7.1 A third-order (3-mode) tensor of dimensions, $I \times J \times K$

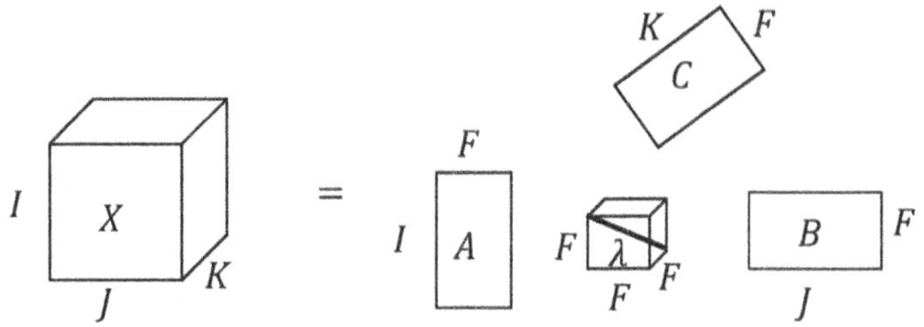

Fig. 7.2 CP decomposition of a 3-mode tensor results in a diagonal core and three factors

7.2.2 *Tensor Decomposition*

Tensor-based algorithms, most notably tensor decomposition, are increasingly important tools for analysis, including clustering, of high-dimensional data sets. Intuitively, tensor decomposition process generalizes matrix decomposition-based data analysis and clustering (such as PCA [7] and SVD [5, 8]) to high-dimensional arrays (known as tensors) and rewrites the given tensor in the form of a set of factor matrices (one for each mode of the input tensor) and a core tensor (which, intuitively, describes the spectral structure of the given tensor). These factor matrices and core tensors then can be used for obtaining multi-modal clusters of the input data. The two most popular tensor decomposition algorithms are the Tucker [31] and the CANDECOMP/PARAFAC(CP) [10] decompositions. We next provide a brief description of these algorithms.

7.2.2.1 CP and Tucker Decompositions

The PARAFAC decomposition can be seen as a generalization of matrix factorizations to tensors [10]. PARAFAC decomposition is also known as CANDECOMP/PARAFAC (CP) decomposition. As shown in Fig. 7.2, given a tensor $\mathcal{X}$, CP factorizes the tensor into F component matrices (where F is a user supplied non-zero integer value also referred to as the *rank* of the decomposition). For the simplicity of the discussion, let us consider a 3-mode tensor $\mathcal{X} \in \mathbb{R}^{I \times J \times K}$. CP would decompose $\mathcal{X}$ into $\mathring{\mathcal{X}}$ consisting of three matrices **A**, **B**, and **C**, such that

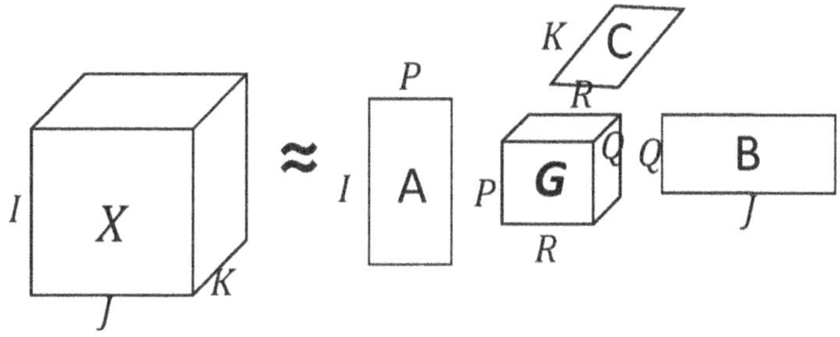

Fig. 7.3 Tucker decomposition of a three-mode tensor

$$\mathcal{X} \approx \tilde{\mathcal{X}} = recombine[\mathring{\mathcal{X}}] \equiv recombine[\mathbf{A}, \mathbf{B}, \mathbf{C}] \equiv \sum_{f=1}^{F} a_f \circ b_f \circ c_f,$$

where $a_f \in \mathbb{R}^I$, $b_f \in \mathbb{R}^J$, and $c_f \in \mathbb{R}^K$. The factor matrices **A**, **B**, **C** are the combinations of the rank-one component vectors into matrices, e.g., $\mathbf{A} = [\, a_1 \; a_2 \cdots a_F \,]$. This is visualized in Fig. 7.2.

Tucker decomposition generalizes singular value matrix decomposition (SVD) to higher-dimensional data (Fig. 7.3). Given a tensor $\mathcal{X} \in \mathbb{R}^{I \times J \times K}$, Tucker decomposition factorizes the tensor into factor matrices with different number of rows, which are referred to as the rank of the decomposition. Tucker decomposition would decompose $\mathcal{X}$ into three matrices **A**, **B**, **C** and one core dense tensor **G**, such that

$$\mathcal{X} \approx \tilde{\mathcal{X}} = \mathbf{G} \times_1 \mathbf{A} \times_2 \mathbf{B} \times_3 \mathbf{C} \equiv \sum_{p=1}^{P} \sum_{q=1}^{Q} \sum_{r=1}^{R} g_{pqr} a_p \circ b_q \circ c_r,$$

where $\mathbf{A} \in \mathbb{R}^{I \times P}$, $\mathbf{B} \in \mathbb{R}^{J \times Q}$, $\mathbf{C} \in \mathbb{R}^{K \times R}$ are the factor matrices and can be treated as the principal components in each mode. The (dense) core tensor, $\mathbf{G} \in \mathbb{R}^{P \times Q \times R}$, indicates the strength of interactions among different components of the factor matrices.

7.2.2.2 Accuracy of Tensor Decomposition

Note that, in general, unlike matrix decomposition (where each matrix has an exact decomposition), tensors may not have exact decompositions [16]. Therefore, many of the algorithms for decomposing tensors are based on an iterative process that tries to improve the approximation until a convergence condition is reached, such as an alternating least squares (ALS) method: at its most basic form, ALS estimates, at each iteration, one factor matrix while maintaining other matrices fixed; this process is repeated for each factor matrix associated with the modes of the input tensor.

Note that due to the approximate nature of tensor decomposition operation, given a decomposition $[\mathbf{A}, \mathbf{B}, \mathbf{C}]$ of $\mathcal{X}$, the tensor $\tilde{\mathcal{X}}$ that one would obtain by re-composing the tensor by combining the factor matrices $\mathbf{A}$, $\mathbf{B}$, and $\mathbf{C}$ is often different from the input tensor, $\mathcal{X}$. The accuracy of the decomposition is often measured by considering the Frobenius norm of the difference tensor:

$$accuracy(\mathcal{X}, \tilde{\mathcal{X}}) = 1 - error(\mathcal{X}, \tilde{\mathcal{X}}) = 1 - \left(\frac{\|\tilde{\mathcal{X}} - \mathcal{X}\|}{\|\mathcal{X}\|} \right).$$

7.2.3 Tensor Decomposition and Clustering

As we mentioned earlier, intuitively, tensor decomposition process generalizes matrix decomposition to high-dimensional arrays and the resulting factor matrices and core tensors then can be used for obtaining multi-modal clusters of the input data. Indeed, tensor-based representations of data and tensor decompositions (especially the two widely used decompositions CP [10] and Tucker [31]) are proven to be effective in multi-aspect data analysis and clustering. For instance, [22] used tensor decomposition to cluster patients in a health-care setting based on their individual and health profile data, including age, medical history, and diagnostics: in particular, the authors have created a patient information tensor and decomposed this tensor (by nonnegative low-rank approximation methods) to obtain semantic clusters that can be used to characterize patients' records. Davidson et al. [6] applied tensor decomposition to fMRI data to help differentiating healthy and Alzheimer affected individuals. Cao et al. [3] used a similar tensor decomposition-based approach to cluster face images: authors modeled a collection of faces as a tensor and they applied a tensor-based principal component analysis for seeking face clusters. Wu et al. [32] leveraged CP decomposition (solved through stochastic gradient descent) to cluster heterogeneous information networks: each type of object in the network is represented as a different mode of the tensor. Sun et al. [28], on the other hand, has shown that Tucker decomposition can be used for subspace clustering which simultaneously conducts dimensionality reduction and membership representation.

7.2.4 Block-Based Tensor Decomposition

One key challenge with tensor decomposition is its computational complexity: decomposition algorithms have high computational costs and, in particular, incur large memory overheads (also known as the *intermediary data blow-up problem*) and, thus, basic algorithms and naive implementations are not suitable for large problems. HaTen2 [12] focuses on sparse tensors and presents a scalable tensor

Algorithm 1 The outline of the block-based iterative improvement process

Input: original tensor $\mathcal{X}$, partitioning pattern $\mathcal{K}$, and decomposition rank, F
Output: CP tensor decomposition $\hat{\mathcal{X}}$
1. Phase 1: for all $\mathbf{k} \in \mathcal{K}$
    - decompose $\mathcal{X}_{\mathbf{k}}$ into $U_{\mathbf{k}}^{(1)}, U_{\mathbf{k}}^{(2)}, \ldots, U_{\mathbf{k}}^{(N)}$
2. Phase 2: repeat
    a. for each mode $i = 1$ to N
        i. for each modal partition, $k_i = 1$ to K_i,
            A. update $A_{(k_i)}^{(i)}$ using $U_{[*,\ldots,*,k_i,*,\ldots,*]}^{(i)}$, for each block $\mathcal{X}_{[*,\ldots,*,k_i,*,\ldots,*]}$; more specifically,
                - compute $T_{(k_i)}^{(i)}$, which involves the use of $U_{[*,\ldots,*,k_i,*,\ldots,*]}^{(i)}$ (i.e., the mode-i factors of $\mathcal{X}_{[*,\ldots,*,k_i,*,\ldots,*]}$)
                - revise $P_{[*,\ldots,*,k_i,*,\ldots,*]}$ using $U_{[*,\ldots,*,k_i,*,\ldots,*]}^{(i)}$ and $A_{(k_i)}^{(i)}$
                - compute $S_{(k_i)}^{(i)}$ using the above
                - update $A_{(k_i)}^{(i)}$ using the above
                - for each $\mathbf{k} = [*, *, \ldots, k_i, \ldots, *, *]$
                    - update $P_{\mathbf{k}}$ and $Q_{\mathbf{k}}$ using
                    - $U_{\mathbf{k}}^{(i)}$ and $A_{(k_i)}^{(i)}$

    until stopping condition
3. Return $\hat{\mathcal{X}}$

decomposition suite of methods for Tucker and PARAFAC decompositions on the MapReduce framework. TensorDB [15] leverages a chunk-based framework to store and retrieve data, extends array operations to tensor operations, and introduces optimization schemes for in-database tensor decomposition.

One way to deal with this challenge is to partition the tensor and obtain the tensor decomposition leveraging these smaller partitions. Block-based decomposition techniques partition the given tensor into blocks or sub-tensors, initially decompose each block independently, and then iteratively combine these decompositions into a final decomposition. GridPARAFAC [23], for example, partitions the tensor into pieces, obtains decomposition for each piece (potentially in parallel), and stitches the partial decomposition results into a combined decomposition for the initial tensor through an iterative improvement process. Here, we provide an overview of the block-based tensor decomposition process.

Let us consider an N-mode tensor $\mathcal{X} \in \mathbb{R}^{I_1 \times I_2 \times \ldots \times I_N}$, partitioned into a set (or grid) of sub-tensors $\mathfrak{X} = \{\mathcal{X}_{\mathbf{k}} \mid \mathbf{k} \in \mathcal{K}\}$ where $\mathcal{K}$ is the set of sub-tensor indexes. Without loss of generality, let us assume that $\mathcal{K}$ partitions the mode i into K_i equal partitions, i.e., $|\mathcal{K}| = \prod_{i=1}^{N} K_i$. Let us also assume that we are given a target decomposition rank, F, for the tensor $\mathcal{X}$. Let us further assume that each sub-tensor in $\mathfrak{X}$ has already been decomposed with target rank F and let $\mathfrak{U}^{(i)} = \{U_{\mathbf{k}}^{(i)} \mid \mathbf{k} \in \mathcal{K}\}$

denote the set of F-rank sub-factors[1] corresponding to the sub-tensors in $\mathfrak{X}$ along mode i. In other words, for each $\mathcal{X}_{\mathbf{k}}$, we have

$$\mathcal{X}_{\mathbf{k}} \approx I \times_1 U_{\mathbf{k}}^{(1)} \times_2 U_{\mathbf{k}}^{(2)} \cdots \times_N U_{\mathbf{k}}^{(N)}, \tag{7.1}$$

where I is the N-mode $F \times F \times \ldots \times F$ identity tensor, where the diagonal entries are all 1s and the rest are all 0s. Given these, Phan and Cichocki [23] presents an iterative improvement algorithm for composing these initial sub-factors into the full F-rank factors, $A^{(i)}$ (each one along one mode), for the input tensor, $\mathcal{X}$. The outline of this block- based process is as follows: Let us partition each factor $A^{(i)}$ into K_i parts corresponding to the block boundaries along mode i:

$$A^{(i)} = [A_{(1)}^{(i)T} A_{(2)}^{(i)T} ... A_{(K_i)}^{(i)T}]^T.$$

Given this partitioning, each sub-tensor $\mathcal{X}_{\mathbf{k}}$, $\mathbf{k} = [k_1, \ldots, k_i, \ldots, k_N] \in \mathcal{K}$ can be described in terms of these sub-factors:

$$\mathcal{X}_{\mathbf{k}} \approx I \times_1 A_{(k_1)}^{(1)} \times_2 A_{(k_2)}^{(2)} \cdots \times_N A_{(k_N)}^{(N)} \tag{7.2}$$

Moreover [23] shows that the current estimate of the sub-factor $A_{(k_i)}^{(i)}$ can be revised using the update rule (for more details on the update rules please see [23]):

$$A_{(k_i)}^{(i)} \longleftarrow \quad T_{(k_i)}^{(i)} \left(S_{(k_i)}^{(i)}\right)^{-1} \tag{7.3}$$

where

$$T_{(k_i)}^{(i)} \quad = \quad \sum_{\mathbf{l} \in \{[*,\ldots,*,k_i,*,\ldots,*]\}} U_{\mathbf{l}}^{(i)} \left(P_{\mathbf{l}} \oslash (U_{\mathbf{l}}^{(i)T} A_{(k_i)}^{(i)})\right)$$

$$S_{(k_i)}^{(i)} \quad = \quad \sum_{\mathbf{l} \in \{[*,\ldots,*,k_i,*,\ldots,*]\}} Q_{\mathbf{l}} \oslash \left(A_{(k_i)}^{(i)T} A_{(k_i)}^{(i)}\right)$$

such that, given $\mathbf{l} = [l_1, l_2, \ldots, l_N]$, we have

- $P_{\mathbf{l}} = \circledast_{h=1}^{N} (U_{\mathbf{l}}^{(h)T} A_{(l_h)}^{(h)})$ and $Q_{\mathbf{l}} = \circledast_{h=1}^{N} (A_{(l_h)}^{(h)T} A_{(l_h)}^{(h)})$.

Above, $\circledast$ denotes the Hadamard product and $\oslash$ denotes element-wise division.

The block-based tensor decomposition process is outlined in pseudocode in Algorithm 1. Figure 7.4 provides a visual example of this process: The given input tensor $\mathcal{X}$ is partitioned into two sub-tensors, $\mathcal{X}_1$ and $\mathcal{X}_2$. In the first stage, each sub-tensor is decomposed by CP, thus obtaining partial factors. The second stage

[1]If the sub-tensor is empty, then the factors are 0 matrices of the appropriate size.

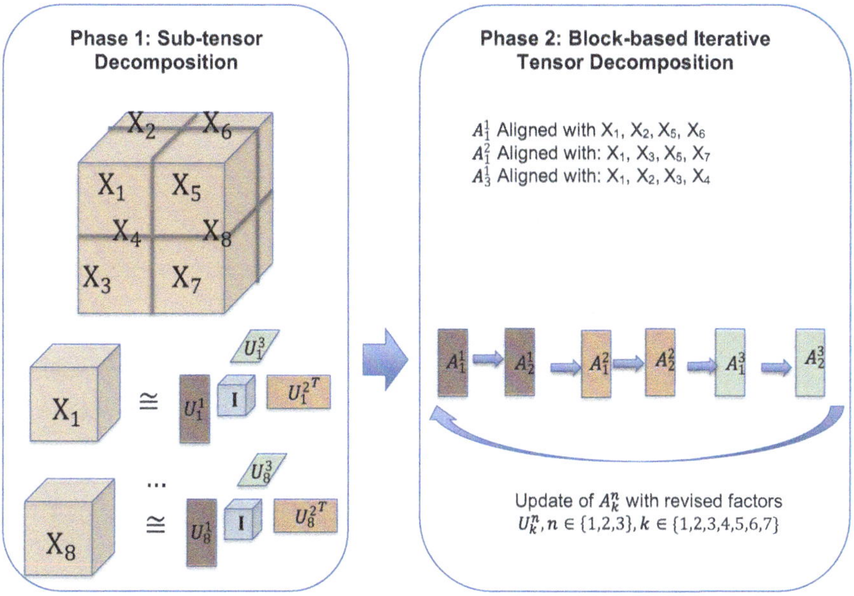

Fig. 7.4 Illustration of block-based tensor decomposition process

combines these partial decomposed factors using iterative updates to derive the final factors (and the corresponding core) for tensor $\mathcal{X}$.

7.3 Sub-Tensor Impact Graphs (SIGs) and Sub-Tensor Impact Scores

In this section, we formally introduce the concept of *sub-tensor impact graph (SIG)* that captures and represents the underlying structure of sub-tensors and helps efficiently calculate the impact of each sub-tensor on the decomposition accuracy of the overall tensor.

Let an N-mode tensor, $\mathcal{X} \in \mathbb{R}^{I_1 \times I_2 \times \ldots \times I_N}$, be partitioned into a grid, $\mathfrak{X} = \{\mathcal{X}_{\mathbf{k}} \mid \mathbf{k} \in \mathcal{K}\}$, of sub-tensors, such that

- K_i indicates the number of partitions along mode-i,
- the size of the jth partition along mode i is $I_{j,i}$ (i.e., $\sum_{j=1}^{K_i} I_{j,i} = I_i$), and
- $\mathcal{K} = \{[k_{j_1}, \ldots, k_{j_i}, \ldots, k_{j_N}] \mid 1 \leq i \leq N,\ 1 \leq j_i \leq K_i\}$ is a set of sub-tensor indexes.

The number, $\|\mathfrak{X}\|$, of partitions (and thus also the number, $\|\mathcal{K}\|$, of partition indexes) is $\prod_{i=1}^{N} K_i$.

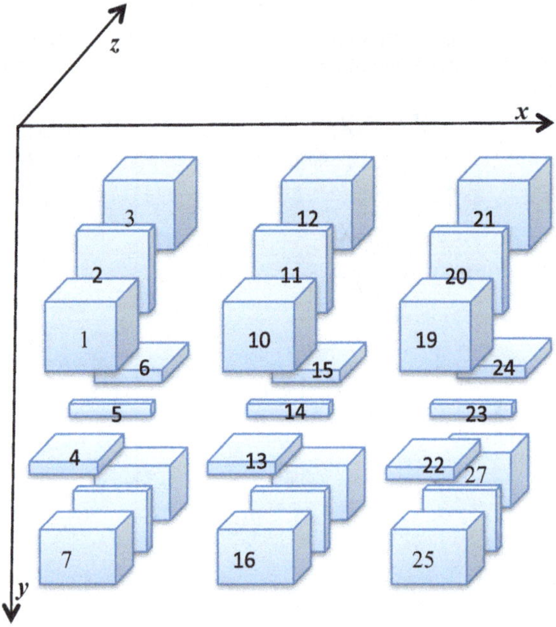

Fig. 7.5 A sample 3-mode tensor, partitioned into 27 heterogeneous sub-tensors

Example 7.1 Figure 7.5 shows a 3-mode tensor, partitioned into 27 sub-tensors: 12 tensor-blocks (sub-tensors 1, 3, 7, 9, 10, 12, 16, 18, 19, 21, 15, 27), 12 slices (sub-tensors 2, 8, 11, 17, 20, 26, 4, 6, 13, 15, 22, 24), and three fibers (sub-tensors 5, 14, 23). The specific shapes of partitions may correspond to user's requirement such as the degree of importance or user focus.

7.3.1 Accuracy Dependency Among Sub-Tensors

In Sect. 7.2.4, we presented update rules block-based tensor decomposition algorithms use for stitching the individual sub-tensor decompositions into a complete decomposition for the whole tensor. While the precise derivation of these update rules is not critical for our discussion (and is beyond the scope of this chapter), it is **important** to note that, as visualized in Fig. 7.6, each $A^{(i)}_{(k_i)}$ is maintained incrementally by using, for all $1 \leq j \leq N$, the current estimates for $A^{(j)}_{(k_j)}$ and the decompositions in $\mathfrak{U}^{(j)}$, i.e., the F-rank sub-factors of the sub-tensors in $\mathfrak{X}$ along the different modes of the tensor. Moreover, and most importantly for the present discussion, this update rule for $A^{(i)}_{(k_i)}$ supports the following observation: Given

$$\mathcal{X}_{\mathbf{k}} \approx I \times_1 A^{(1)}_{(k_1)} \times_2 A^{(2)}_{(k_2)} \cdots \times_N A^{(N)}_{(k_N)},$$

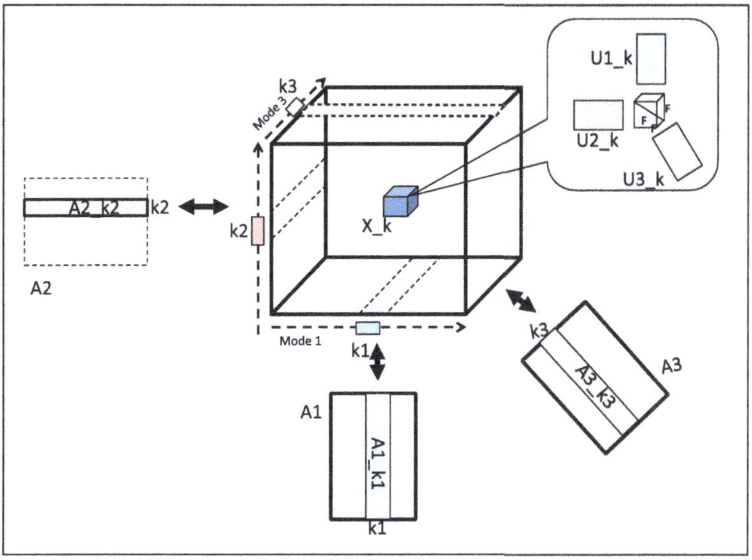

Fig. 7.6 The block-based update rule maintains $A^{(i)}_{(k_i)}$ incrementally by using the current estimates for $A^{(j)}_{(k_j)}$ and the decompositions in $\mathfrak{U}^{(j)}$

the final accuracy for the sub-tensor $\mathcal{X}_{\mathbf{k}}$, $\mathbf{k} = [k_1, \ldots, k_i, \ldots, k_N]$, depends on the accuracies of sub-factors $A^{(i)}_{(k_i)}$. Moreover, the accuracy of each of these, in turn, depends on the accuracies of the sub-factors of the contributing sub-tensors. More specifically, when updating $A^{(i)}_{(k_i)}$, we need to compute

- $T^{(i)}_{(k_i)}$, which involves the use of $U^{(i)}_{[*,\ldots,*,k_i,*,\ldots,*]}$ (i.e., the mode-i factors of $\mathcal{X}_{[*,\ldots,*,k_i,*,\ldots,*]}$), and
- $P_{[*,\ldots,*,k_i,*,\ldots,*]}$, which in turn uses $U^{(h)}_{[*,\ldots,*,k_i,*,\ldots,*]}$ for $1 \le h \le N$ (i.e., all factors of $\mathcal{X}_{[*,\ldots,*,k_i,*,\ldots,*]}$).

Therefore, the final accuracy of $\mathcal{X}_{\mathbf{k}}$ depends directly on the initial decomposition accuracies of the factor matrices $U^{(h)}_{[*,\ldots,*,k_i,*,\ldots,*]}$, for $1 \le i, h \le N$.

In other words, for each sub-tensor $\mathcal{X}_{\mathbf{k}}$, there is a set, $direct\_impact(\mathcal{X}_{\mathbf{k}}) \subseteq \mathfrak{X}$, of sub-tensors that consists of those sub-tensors whose initial decomposition accuracies directly impact the final decomposition accuracy of $\mathcal{X}_{\mathbf{k}}$. Moreover, as visualized in Fig. 7.7, $direct\_impact(\mathcal{X}_{\mathbf{k}})$ consists of those sub-tensors that are aligned (i.e., share the same slices) with $\mathcal{X}_{\mathbf{k}}$, along the different modes of the tensor.

7.3.2 *Sub-Tensor Impact Graphs (SIGs)*

Given the accuracy dependencies among the sub-tensors formalized above, we can define a *sub-tensor impact graph (SIG)*:

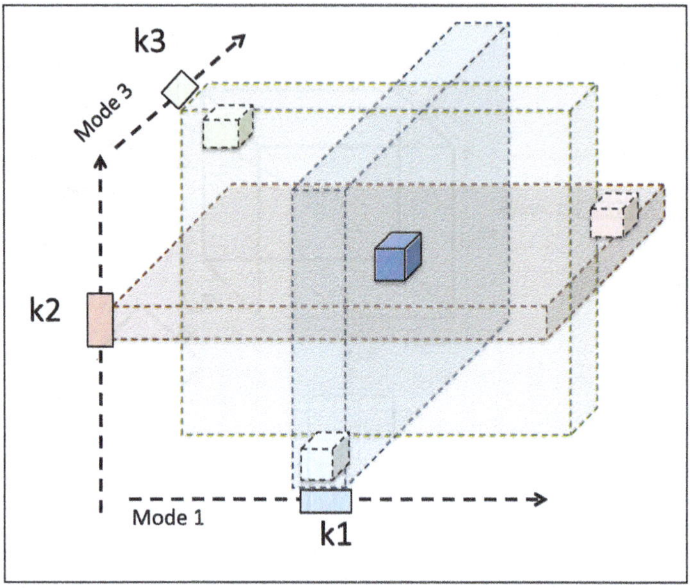

Fig. 7.7 The sub-tensors whose initial decomposition accuracies directly impact given sub-tensors are aligned (i.e., share the same slices) with that sub-tensor along the different modes of the tensor

Definition 7.1 Let an N-mode tensor, $\mathcal{X} \in \mathbb{R}^{I_1 \times I_2 \times \ldots \times I_N}$, be partitioned into a grid, $\mathfrak{X} = \{\mathcal{X}_{\mathbf{k}} \mid \mathbf{k} \in \mathcal{K}\}$, of sub-tensors. The corresponding *sub-tensor impact graph (SIG)* is a directed, weighted graph, $G(V, E, w())$, where

- for each $\mathcal{X}_{\mathbf{k}} \in \mathfrak{X}$, there exists a corresponding $v_{\mathbf{k}} \in V$,
- for each $\mathcal{X}_{\mathbf{l}} \in direct\_impact(\mathcal{X}_{\mathbf{k}})$, there exists a directed edge $v_{\mathbf{l}} \rightarrow v_{\mathbf{k}}$ in E, and
- $w()$ is an edge weight function, such that $w(v_{\mathbf{l}} \rightarrow v_{\mathbf{k}})$ quantifies the direct accuracy impact of decomposition accuracy of $\mathcal{X}_{\mathbf{l}}$ on $\mathcal{X}_{\mathbf{k}}$. ◊

Intuitively, the sub-tensor impact graph represents how the decomposition accuracies of the given set of sub-tensors of an input tensor impact the overall combined decomposition accuracy. A key requirement, of course, is to define the edge weight function, $w()$, that quantifies the accuracy impacts of the sub-tensors that are related through update rules. In this section, we introduce three alternative strategies to account for the propagation of impacts within the tensor during the decomposition process.

7.3.2.1 Alt. #1: Uniform Edge Weights

The most straightforward way to set the weights of the edges in E is to assume that the propagation of the inaccuracies over the sub-tensor impact graph is uniform. In other words, in this case, for all $e \in E$, we set $w_{\text{uni}}(e) = 1$.

7.3.2.2 Alt. #2: Surface of Interaction-Based Weights

While being simple, the uniform edge weight alternative may not properly account for the impact of the varying dimensions of the sub-tensors on the error propagation.

As we see in Fig. 7.5, in general, the neighbors of a given sub-tensor can be of varying shape and dimensions and we may need to account for this diversity in order to properly assess how inaccuracies propagate in the tensor. In particular, in this subsection, we argue that the *surface of interaction* between two sub-tensors $\mathcal{X}_{\mathbf{j}}$ and $\mathcal{X}_{\mathbf{l}}$, defined as below, may need to be considered to account for impact propagation:

Definition 7.2 (Surface of Interaction) Let $\mathcal{X}$ be a tensor partitioned into a set (or grid) of sub-tensors $\mathfrak{X} = \{\mathcal{X}_{\mathbf{k}} \mid \mathbf{k} \in \mathcal{K}\}$. Let also $\mathcal{X}_{\mathbf{j}}$ and $\mathcal{X}_{\mathbf{l}}$ be two sub-tensors in $\mathfrak{X}$, such that

- $\mathbf{j} = [k_{j_1}, k_{j_2}, \ldots, k_{j_N}]$ and
- $\mathbf{l} = [k_{l_1}, k_{l_2}, \ldots, k_{l_N}]$.

We define the *surface of interaction*, $surf(\mathcal{X}_{\mathbf{j}}, \mathcal{X}_{\mathbf{l}})$, between $\mathcal{X}_{\mathbf{j}}$ and $\mathcal{X}_{\mathbf{l}}$ as follows:

$$surf(\mathcal{X}_{\mathbf{j}}, \mathcal{X}_{\mathbf{l}}) = \prod_{h \ s.t.\ j_h = l_h} I_{j_h,h}.$$

◊

Here $I_{j_h,h}$ is the size of the j_hth partition along mode h.

Principle 1 *Let $G(V, E, w())$ be a sub-tensor impact graph and let $(v_{\mathbf{j}} \rightarrow v_{\mathbf{l}}) \in E$ be an edge in the graph. The weight of this edge from $v_{\mathbf{j}}$ to $v_{\mathbf{l}}$ should reflect the area of the surface of interaction between the sub-tensors $\mathcal{X}_{\mathbf{j}}$ and $\mathcal{X}_{\mathbf{l}}$.*

Intuitively, this principle verbalizes the observation that impacts are likely to propagate more easily if two sub-tensors share large dimensions along the modes on which their partitions coincide. Under this principle, we can set the weight of the edge $(v_{\mathbf{j}} \rightarrow v_{\mathbf{l}}) \in E$ as follows:

$$w_{\text{sur}}(v_{\mathbf{j}} \rightarrow v_{\mathbf{l}}) = \frac{surf(\mathcal{X}_{\mathbf{j}}, \mathcal{X}_{\mathbf{l}})}{\sum_{(v_{\mathbf{j}} \rightarrow v_{\mathbf{m}}) \in E} surf(\mathcal{X}_{\mathbf{j}}, \mathcal{X}_{\mathbf{m}})}.$$

7.3.2.3 Alt. #3: Value Alignment-Based Edge Weights

Although surface of interaction-based edge weights can potentially account for the varying shapes and sizes of the sub-tensors of $\mathcal{X}$, they fail to take into account for how similar these sub-tensors are—more specifically, they ignore how the values within the sub-tensors are distributed and whether these distributions are aligned across them.

Intuitively, if the value distributions are aligned (or similar) along the modes that two sub-tensors share, then they are likely to have high impacts on each other's decomposition during the decomposition process. If they are dissimilar, on the other hand, their impacts on each other will be minimal. Therefore, considering only the area of the surface of interaction may not be sufficient to properly account for the inaccuracy propagation within the tensor. More specifically, we need to measure the value alignment between sub-tensors as well:

Definition 7.3 (Value Alignment) Let $\mathcal{X}$ be a tensor partitioned into a set (or grid) of sub-tensors $\mathfrak{X} = \{\mathcal{X}_{\mathbf{k}} \mid \mathbf{k} \in \mathcal{K}\}$. Let also $\mathcal{X}_{\mathbf{j}}$ and $\mathcal{X}_{\mathbf{l}}$ be two sub-tensors in $\mathfrak{X}$, such that

- $\mathbf{j} = [k_{j_1}, k_{j_2}, \ldots, k_{j_N}]$ and
- $\mathbf{l} = [k_{l_1}, k_{l_2}, \ldots, k_{l_N}]$.

Let, $A = \{h \mid k_{j_h} = k_{l_h}\}$ be the set of modes along which the two sub-tensors are aligned and let R be the remaining modes. We define the *value alignment*, $align(\mathcal{X}_{\mathbf{j}}, \mathcal{X}_{\mathbf{l}}, A)$, between $\mathcal{X}_{\mathbf{j}}$ and $\mathcal{X}_{\mathbf{l}}$ as

$$align(\mathcal{X}_{\mathbf{j}}, \mathcal{X}_{\mathbf{l}}, A) = cos(\mathbf{c_j}(A), \mathbf{c_l}(A)),$$

where the vector $\mathbf{c_j}(A)$ is constructed from the sub-tensor $\mathcal{X}_{\mathbf{j}}$ as follows[2]:

$$\mathbf{c_j}(A) = vectorize(\mathcal{M}_{\mathbf{j}}(A))$$

and the tensor $\mathcal{M}_{\mathbf{j}}(A)$ is constructed from $\mathcal{X}_{\mathbf{j}}$ by fixing the values along the modes in A: $\forall 1 \leq i_h \leq I_{j_h,h}$,

$$\mathcal{M}_{\mathbf{j}}(A)[i_1, i_2, \ldots, i_{|A|}] = norm(\mathcal{X}_{\mathbf{j}}|_{A,i_1,i_2,\ldots,i_{|A|}}).$$

Here, $norm()$ is the standard Frobenius norm and $\mathcal{X}_{\mathbf{j}}|_{A,i_1,i_2,\ldots,i_{|A|}}$ denotes the part of $\mathcal{X}_{\mathbf{j}}$ where the modes in A take values i_1,i_2, through $i_{|A|}$. ◊

Intuitively, $\mathbf{c_j}(A)$ captures the value distribution of the tensor $\mathcal{X}_{\mathbf{j}}$ along the modes in A.

Principle 2 *Let $G(V, E, w())$ be a sub-tensor impact graph and let $(v_{\mathbf{j}} \rightarrow v_{\mathbf{l}}) \in E$ be an edge in the graph. The weight of this edge from $v_{\mathbf{j}}$ to $v_{\mathbf{l}}$ should reflect the structural alignment between the sub-tensors $\mathcal{X}_{\mathbf{j}}$ and $\mathcal{X}_{\mathbf{l}}$.*

This principle verbalizes the observation that impacts are likely to propagate more easily if two given sub-tensors are structurally aligned along the modes on which their partitions coincide. As before, under this principle, we can set the edge weights of the edge $(v_{\mathbf{j}} \rightarrow v_{\mathbf{l}}) \in E$ in the sub-tensor impact graph as follows:

$$w_{\text{align}}(v_{\mathbf{j}} \rightarrow v_{\mathbf{l}}) = \frac{align(\mathcal{X}_{\mathbf{j}}, \mathcal{X}_{\mathbf{l}})}{\sum_{(v_{\mathbf{j}} \rightarrow v_{\mathbf{m}}) \in E} align(\mathcal{X}_{\mathbf{j}}, \mathcal{X}_{\mathbf{m}})}.$$

[2] $\mathbf{c_l}(A)$ is similarly constructed from sub-tensor $\mathcal{X}_{\mathbf{l}}$.

7.3.2.4 Alt. #4: Combined Edge Weights

The surface of interaction-based edge weights account for the shapes of the sub-tensors, but do not account for their value alignments. In contrast, value alignment-based edge weights consider the structural similarities of the sub-tensors, but ignore how big the surfaces they share are.

Therefore, a potentially more effective alternative would be to combine these surface of interaction and value alignment-based edge weights into a single weight that takes into account both aspects of sub-tensor interaction:

$$w_{\text{comb}}(v_{\mathbf{j}} \rightarrow v_{\mathbf{l}}) = \frac{comb(\mathcal{X}_{\mathbf{j}}, \mathcal{X}_{\mathbf{l}})}{\sum_{(v_{\mathbf{j}} \rightarrow v_{\mathbf{m}}) \in E} comb(\mathcal{X}_{\mathbf{j}}, \mathcal{X}_{\mathbf{m}})},$$

where $comb(\mathcal{Y}, \mathcal{Z}) = align(\mathcal{Y}, \mathcal{Z}) \times surf(\mathcal{Y}, \mathcal{Z})$.

7.3.3 Sub-Tensor Impact Scores

While the edges on the sub-tensor impact graph, G, account for how (in)accuracies propagate during each individual application of the update rules, it is important to note that after several iterations of updates, *indirect* propagation of impacts also occur over the graph G:

- during the first application of the update rule, impacts propagate among the sub-tensors that are immediate neighbors;
- during the second application of the update rule, impacts reach from one sub-tensor to those sub-tensors that are 2-hop away;

  ...
- during the mth application of the rule, impacts propagate to the m-hop neighbors of each sub-tensor.

In order to use the sub-tensor impact graph to assign resources, we therefore need to measure how impacts propagate within G over a large number of iterations of the alternating least squares (ALS) process.

For this purpose, we rely on a random-walk-based measure of node relatedness on the given graph. More specifically, we rely on personalized PageRank (PPR [2, 4]) to measure sub-tensor relatedness. Like all random-walk-based techniques, PPR encodes the structure of the graph in the form of a transition matrix of a stochastic process and complements this with a seed node set, $S \subseteq V$, which serves as the context in which scores are assigned: each node, v_i in the graph is associated with a score based on its positions in the graph relative to this seed set (i.e., how many paths there are between v_i and the seed set and how short these paths are). Intuitively, these seeds represent sub-tensors that are critical in the given application (e.g. high-update, high-noise, or high-user-relevance; see Sects. 7.3 through 7.3.2 for various applications).

Given the graph and the seeds, the PPR score $\mathbf{p}[i]$ of v_i is obtained by solving the following equation:

$$\mathbf{p} = (1-\beta)\mathbf{T}_G\,\mathbf{p} + \beta\mathbf{s},$$

where $\mathbf{T}_G$ denotes the transition matrix corresponding to the graph G (and the underlying edge weights) and $\mathbf{s}$ is a re-seeding vector such that if $v_i \in S$, then $\mathbf{s}[i] = \frac{1}{\|S\|}$ and $\mathbf{s}[i] = 0$, otherwise. Intuitively, $\mathbf{p}$ is the stationary distribution of a random walk on G which follows graph edges (according to the transition probabilities $\mathbf{T}_G$) with probability $(1-\beta)$ and jumps to one of the seeds with probability β. Correspondingly, those nodes that are close to the seed nodes over a large number of paths obtain large scores, whereas those that are poorly connected to the nodes in S receive small PPR scores. We note that the iterative nature of the random-walk process underlying PPR fits well with how inaccuracies propagate during the iterative ALS process. Based on this observation, given a directed, weighted *sub-tensor impact graph (SIG)*, $G(V, E, w())$, we construct a transition matrix, $\mathbf{T}_G$, and obtain the PPR score vector $\mathbf{p}$ by solving the above equation.[3] The resulting *sub-tensor impact scores* are then used for assigning appropriate resources to the various sub-tensors as described in the next three sections.

7.4 Application #1: Block-Incremental CP Decomposition (BICP) and Update Scheduling Based on Sub-Tensor Impact Scores

There are many applications in which data is evolving dynamically. Obviously, in such scenarios, re-computation of the whole tensor decomposition with each update will cause high computational costs. In this section, we present a block-incremental CP decomposition (BICP) scheme which leverages SIGs to efficiently conduct the iterative refinement process during the second phase of the block-based tensor decomposition process. Let us assume that we are given a tensor, $\mathcal{X}$, with decomposition, $\mathring{\mathcal{X}}$, and an update, Δ, on the tensor. BICP significantly reduces computational cost of obtaining the decomposition of the updated tensor, while maintaining high accuracy by relying on two complementary techniques:

- **Update-Sensitive Block Maintenance in First Phase:** In its first phase of the process, instead of repeatedly conducting ALS on each sub-tensor, BICP only revises the decompositions of the sub-tensors that contain updated data. Moreover, when the update is small with respect to the block size, BICP relies on incremental factor tracking [20, 27] to avoid re-decomposition of the updated sub-tensor.

[3]Note that, since in general, the number of partitions is small and is independent of the size of the input tensor, the cost of the PPR computation to obtain the ranks is negligible next to the cost of tensor decomposition.

- **Update-Sensitive Refinement in the Second Phase:** In its second phase, BICP leverages (automatically extracted) metadata about how decompositions of the sub-tensors impact each other's decompositions and a block-centric iterative refinement to help achieve high efficiency and accuracy:
  - BICP limits the refinement process to only those blocks that are aligned with the updated block.
  - We employ sub-tensor impact graph (SIG) to account for the refinement relationships among the sub-tensors; we further apply impact score to reduce redundant work: we
    - identify sub-tensors that do not need to be refined and (probabilistically) prune them from further consideration, and/or
    - assign different ranks to different sub-tensors according to their impact score: naturally, the larger the impact likelihood of a sub-tensor is, the larger target rank BICP assigns to that tensor.

  Intuitively, the above process enables BICP to assign appropriate levels of accuracy to sub-tensors in a way that reflects the distribution of the updates on the whole tensor. This ensures that the process is fast and accurate.

In this chapter, we focus on the SIG-based update sensitive refinement during the second phase of the block-based decomposition process.

7.4.1 Reducing Redundant Refinements

During the refinement process of Phase 2, those sub-tensors that have direct refinement relationships with the updated sub-tensors are critical to the refinement process. Our key observation is that if we could quantify how much an update on a sub-tensor impacts sub-factors on other sub-tensors, then we could use this to optimize Phase 2. More specifically, given an update, Δ on tensor $\mathcal{X}$, BICP assigns an update sensitive impact score, $I_\Delta(\mathcal{X}_\mathbf{k})$ to each sub-tensor, $\mathcal{X}_\mathbf{k}$, and leverages this impact score to regulate the refinement process to eliminate redundant work

Intuitively, if the two sub-tensors are similarly distributed along the modes that they share, then they are likely to have high impacts on each other's decomposition; Therefore we use alternative #3: value alignment-based edge weights to assign the weight of edge (introduced in Sect. 7.3.2). To calculate an update sensitive impact score, we can rely on personalized PageRank (introduced in Sect. 7.3.3) to measure sub-tensor relatedness. PPR encodes the structure of the graph in the form of a transition matrix of a stochastic process from which the significances of the nodes in the graph can be inferred. Here, we choose updated sub-tensors as seed nodes and calculate PPR scores for all the other nodes as their impact scores.

- *Optimization Phase 2-I:* Intuitively, if a sub-tensor has a low impact score, its decomposition is minimally affected given the update, Δ. Therefore, those sub-tensors with very low-impact factors can be completely ignored in the refinement process and their sub-factors can be left as they are without any refinement.
- *Optimization Phase 2-P:* While optimization phase 2-I can potentially save a lot of redundant work, completely ignoring low-impact tensors may have a significant impact on accuracy. An alternative approach, with a less drastic impact than ignoring sub-tensors, is to associate a refinement probability to sub-tensors based on their impact scores. In particular, instead of completely ignoring those sub-tensors with low-impact factors, we assign them an update probability, $0 < prob\_update < 1$. Consequently, while the factors of sub-tensors with high impact scores are refined at every iteration of the refinement process, factors of sub-tensors with low-impact scores have lesser probabilities of refinement and, thus, do not get refined at every iteration of Phase 2.
- *Optimization Phase 2-R:* A second alternative to completely ignoring the refinement process for low-impact sub-tensors is to assign different ranks to different sub-tensors according to their impact scores: naturally, the higher the target rank is, the more accurate the decomposition of the sub-tensor is. We achieve this by adjusting the decomposition rank, $F_{\mathbf{k}}$ of $\mathcal{X}_{\mathbf{k}}$, as a function of the corresponding tensor's update sensitive impact score:

$$F_{\mathbf{k}} = \left\lceil F \times \frac{I_{\delta}(\mathcal{X}_{\mathbf{k}})}{max_{\mathbf{h}}\{I_{\delta}(\mathcal{X}_{\mathbf{h}})\}} \right\rceil .$$

Intuitively, this formula sets the decomposition rank of the sub-tensor with the highest impact score relative to the given update, Δ, to F; other sub-tensors are assigned progressively smaller ranks (potentially all the way down to 1)[4] based on their impacts scores. Once the new ranks are computed, we obtain new $U_{(k)}$ factors with partial ranks $F_{\mathbf{k}}$ for $\mathcal{X}_{\mathbf{k}}$ and refine these incrementally in Phase 2.

Here, we consider two rank-based optimization strategies, *phase 2-Ra* and *phase 2-Ri*. In *phase 2-Ra*, we potentially adjust the decomposition rank for all relevant sub-tensors. In *phase 2-Ri*, however, we adjust ranks only for sub-tensors with high impact on the overall decomposition.

By extending the complexity formulation from [23], we can obtain the complexity[5] of Phase 2 as $O((F \times \sum_{i=1}^{N} \frac{I_i}{K_i} + F^2) \times \mathcal{T} \times H \times |\mathcal{D}|)$ where $\mathcal{T}$ is the number of refinement iterations, $H = (100 - L)\%$ is the ratio of high impact sub-tensors maintained, $|\mathcal{D}|$ is the number of sub-tensors that have direct impact on updated sub-tensors, I_i is the dimensionality of the tensor along mode i, and K_i is the number of partitions along that mode.

---

[4]It is trivial to modify this equation such that the smallest rank will correspond to a user provided lower bound, $F_{\min}$, when such a lower bound is provided by the user.

[5]Here we report the complexity of $phase2 - I$ and other refinement method complexity can be derived similarly.

7.4.2 *Evaluation*

In this section, we report sample results that aim to assess the effectiveness of the proposed BICP approach in helping eliminate redundant refinements

7.4.2.1 Setup

Data Sets In these experiments, we used three data sets: *Epinions* [29], *Ciao* [29], and *Enron* [24]. The first two of these are comparable in terms of their sizes and semantics: they are both $5000 \times 5000 \times 27$ tensors, with schema $\langle user, item, category\rangle$, and densities 1.089×10^{-6} and 1.06×10^{-6}, respectively. The *Enron* email data set, on the other hand, has dimensions $5632 \times 184 \times 184$, density 1.8×10^{-4}, and schema $\langle time, from, to\rangle$.

Data Updates We divided the tensor into 64 blocks (using $4 \times 4 \times 4$ partitioning) and applied all the updates to four of these blocks; Once the blocks are selected, we randomly pick a slice on the block and update 10% of the fibers on this slice.

Alternative Strategies We consider the following strategies to maintain the tensor decomposition: Firstly, we apply the basic two-phase block-centric decomposition strategy, i.e., we decompose all sub-tensors with CPALS in Phase 1 and we apply iterative refinement using all sub-tensors in Phase 2 (in the charts, we refer to this non-incremental decomposition approach as `ORI`). For Phase 1, we use a version of STA where we update fibers that are update-critical, i.e., with highest energy among all the affected fibers. For Phase 2, again, we have several alternatives: (a) applying Phase 2 without any impact score-based optimization (`P2N`), (b) ignoring $L\%$ of sub-tensors with the lowest impact scores (`P2I`), (c) reducing the decomposition rank of sub-tensors (`P2Ra` and `P2Ri`), or (d) using probabilistic refinements for sub-tensors with low impact scores (`P2P`). In these experiments, we choose $L = 50\%$ and, for `P2P`, we set the update probability to $p = 0.1$. In addition to the block-based BCIP and its optimizations, we also considered, as an efficient alternative, application of the incremental factor tracking process to the whole tensor as in STA [27]—in the charts, we refer to this alternative approach as `Whole`.

Evaluation Criteria We use the measure reported in Sect. 7.2.2.1 to assess decomposition accuracy. We also report decomposition time for different settings. In these experiments, the target decomposition rank is set to $F = 10$. Unless otherwise specified, the maximum number of iterations in Phase 2 is set to 1000. Each experiment was run 100 times and averages are reported.

Hardware and Software We used a quad-core Intel(R) Core(TM)i5-2400 CPU @ 3.10GHz machine with 8.00GB RAM. All codes were implemented in Matlab and run using Matlab 7.11.0 (2010b) and Tensor Toolbox Version 2.5 [1].

7.4.2.2 Discussion of the Results

Impact scores measure how different regions of the tensor impacts other parts of the tensor during the alternating least squares (ALS) process. Therefore, we expect that, when we leverage the impact scores (computed in a way to account for the distribution of the data updates) to assign the decomposition ranks, we should be able to focus the decomposition work to better fit the dynamically evolving data. Figure 7.8 compares execution times and accuracies of several approaches. Here,

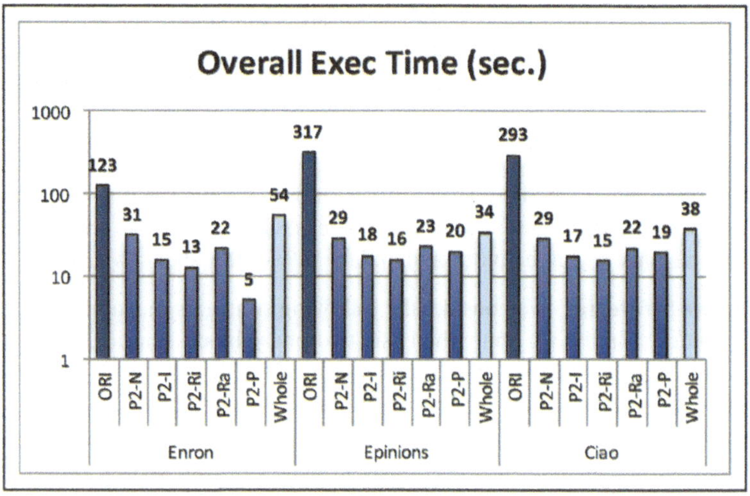

(a)

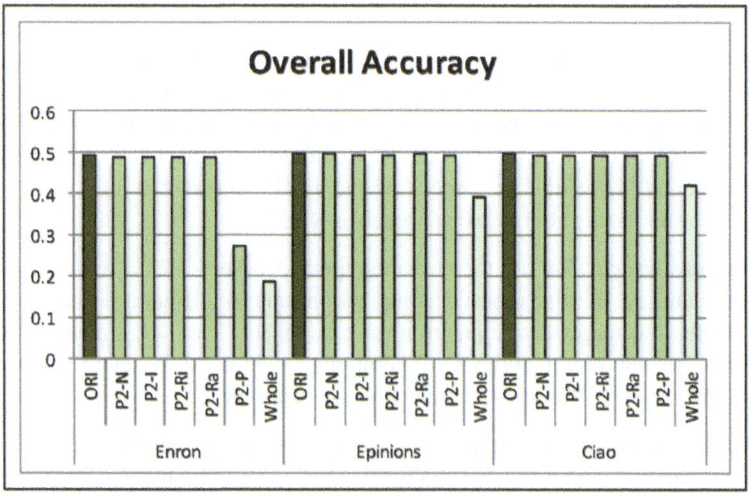

(b)

Fig. 7.8 Comparison of (**a**) Execution times and (**b**) Decomposition accuracies under the default configuration: the proposed optimizations provide several orders of gain in execution time relative to `ORI`, while (unlike `Whole`) they match `ORI`'s accuracy

`ORI` indicates non-incremental two-phase block centric decomposition, whereas `Whole` indicates application of factor tracking to the whole tensor. The other five techniques in the figure (`P2N`, `P2I`, `P2Ri`, `P2Ra`, `P2P`) all correspond to optimizations of the proposed BICP approach for Phase 2.

Firstly, this figure shows that the two social media data sets, Epinions and Ciao, with similar sizes and densities show very similar execution time and accuracy patterns. The figure also shows that the Enron data set also exhibits a pattern roughly similar to the other data sets, despite having a different size and density.

The key observation in Fig. 7.8 is that the SIG-based optimizations provide several orders of gain in execution time while matching the accuracy of non-optimized version almost perfectly (i.e., the optimizations come without accuracy penalties). In contrast, the alternative strategy, `Whole`, which incrementally maintains the factors of the whole tensor (as opposed to maintaining the factors of its blocks) also provides execution time gains, but sees a significant drop in its accuracy.

We note that `P2P`, which probabilistically updates low-impact sub-tensors rather than completely ignoring them, does not significantly improve accuracy. This is because the `P2I` approach already has an accuracy almost identical to `P2N`, i.e., ignoring low-impact tensors is a very safe and effective method to save redundant work. Therefore, also considering that, unless a large number of blocks are ignored, `P2I` is able to match the accuracy of `P2N`, we do not see a major need to use `P2P` to reduce the impact of ignored sub-tensors.

7.5 Application #2: Noise-Profile Adaptive Decomposition (nTD) and Sample Assignment Based on Sub-Tensor Impact Scores

Many of the tensor decomposition schemes are sensitive to noisy data, an inevitable problem in the real world that can lead to false conclusions. Recent research has shown that it is possible to avoid over-fitting by relying on probabilistic techniques that leverage Gibbs sampling-based Bayesian model learning [33]; however, these assume that all the data and intermediary results can fit in the main memory, and (b) they treat the entire tensor uniformly, ignoring potential non-uniformities in the noise distribution. In this chapter, we present a *Noise Adaptive Tensor Decomposition (*`nTD`*)* method, which leverages a probabilistic two-phase decomposition strategy, complemented with sub-tensor impact graphs, to develop a sample assignment strategy that best suits the noise distribution of the given tensor to leverage available *rough* knowledge regarding where in the tensor noise might be more prevalent.

Algorithm 2 Phase 1: Monte Carlo-based Bayesian decomposition of each sub-tensor

Input: Sub-tensor $\mathcal{X}_{\mathbf{k}}$, sampling number L

Output: Decomposed factors $U_{\mathbf{k}}^{(1)}, U_{\mathbf{k}}^{(2)}, \ldots, U_{\mathbf{k}}^{(N)}$

1. Initialize model parameters $U_{\mathbf{k}}^{(1)1}, U_{\mathbf{k}}^{(2)1}, \ldots, U_{\mathbf{k}}^{(N)1}$.
2. For $l = 1, \ldots, L$

   a. Sample the hyper-parameter, α:

      - $\alpha^l \sim p(\alpha^l | U_{\mathbf{k}}^{(1)l}, U_{\mathbf{k}}^{(2)l}, \ldots, U_{\mathbf{k}}^{(N)l}, \mathcal{X}_{\mathbf{k}})$

   b. For each mode $j = 1, \ldots, N$,

      i. Sample the corresponding hyper-parameter, Θ:

         - $\Theta_{U_{\mathbf{k}}^{(j)l}} \sim p(\Theta_{U_{\mathbf{k}}^{(j)l}} | U_{\mathbf{k}}^{(j)l})$

      ii. For $i_j = 1, \ldots, I_j$, sample the mode (in parallel):

$$U_{\mathbf{k}(i_j)}^{(j)(l+1)} \sim p\left(U_{\mathbf{k}(i_j)}^{(j)} \middle| U_{\mathbf{k}}^{(1)l}, \ldots, U_{\mathbf{k}}^{(j-1)l}, U_{\mathbf{k}}^{(j+1)l}, \ldots, U_{\mathbf{k}}^{(N)l}, \Theta^l_{U_{\mathbf{k}}^{(j)}}, \alpha^l, \mathcal{X}_{\mathbf{k}}\right)$$

3. For each mode $j = 1, \ldots, N$,

   - $U_{\mathbf{k}}^{(j)} = \frac{\sum_{i=1}^{L} U_{\mathbf{k}}^{(j)i}}{L}$

More specifically, nTD partitions the tensor into multiple sub-tensors and then decomposes each sub-tensor probabilistically through Bayesian factorization—the resulting decompositions are then recombined through an iterative refinement process to obtain the decomposition for the whole tensor. We refer to this as *Grid-Based Probabilistic Tensor Decomposition* (GPTD). nTD complements GPTD with a SIG-based resource allocation strategy that accounts for the impact of the noise density of one sub-tensor on the decomposition accuracies of the other sub-tensors. This provides several benefits: Firstly, the partitioning helps ensure that the memory footprint of the decomposition is kept low. Secondly, the probabilistic framework used in the first phase ensures that the decomposition is robust to the presence of noise in the sub-tensors. Thirdly, a priori knowledge about noise distribution among the sub-tensors is used to obtain a resource assignment strategy that best suits the noise profile of the given tensor.

7.5.1 Grid-Based Probabilistic Tensor Decomposition (GPTD)

As a block-based algorithm, *Grid-Based Probabilistic Tensor Decomposition* (GPTD) partitions the given tensor into blocks, decomposes each block independently, and then iteratively combines these decompositions into a final composition. Differently from Algorithm 1, however, GPTD leverages Monte Carlo-based

Bayesian decomposition of sub-tensors in its Phase 1 (see Algorithm 2) to better deal with the problem of over-fitting, which is a challenge especially when the data is noisy.

Intuitively, entries in the factor matrices are modeled as probabilistic variables and decomposition is posed as a maximization problem where these (latent) random variables fit the observed data. In the presence of noise in the data, the observed variables may also be modeled probabilistically: since the observations cannot be precisely described, they may be considered as samples from a probability distribution. In this section, following [25], in the presence of data uncertainty (due to noise), we describe the fit between the observed data and the predicted latent factor matrices, probabilistically, as follows:

$$\mathcal{X}_{\mathbf{k}(i_1,i_2,\ldots,i_N)} \Big| U^{(1)}_{\mathbf{k}}, U^{(2)}_{\mathbf{k}} \ldots, U^{(N)}_{\mathbf{k}} \sim \mathcal{N}([U^{(1)}_{\mathbf{k}(i_1)}, U^{(2)}_{\mathbf{k}(i_2)} \ldots, U^{(N)}_{\mathbf{k}(i_N)}], \alpha^{-1}), \tag{7.4}$$

where the conditional distribution of $\mathcal{X}_{\mathbf{k}(i_1,i_2,\ldots,i_N)}$ given $U^{(j)}_{\mathbf{k}}$ $(1 \le j \le N)$ is a Gaussian distribution with mean $[U^{(1)}_{\mathbf{k}(i_1)}, U^{(2)}_{\mathbf{k}(i_2)}, \ldots, U^{(N)}_{\mathbf{k}(i_N)}]$ and the observation precision α. We also impose independent Gaussian priors on the modes:

$$U^{(j)}_{\mathbf{k}(i_j)} \sim \mathcal{N}(\mu_{U^{(j)}_{\mathbf{k}}}, \Lambda^{-1}_{U^{(j)}_{\mathbf{k}}}) \quad i_j = 1 \ldots I_j \tag{7.5}$$

where I_j is the dimensionality of the jth mode. Given this, one can estimate the latent features $U^{(j)}_{\mathbf{k}}$ by maximizing the logarithm of the posterior distribution, $\log p(U^{(1)}_{\mathbf{k}}, U^{(2)}_{\mathbf{k}} \ldots, U^{(N)}_{\mathbf{k}} | \mathcal{X}_{\mathbf{k}})$.

One difficulty with the approach, however, is the tuning of the hyper-parameters of the model: α and $\Theta_{U^{(j)}_{\mathbf{k}}} \equiv \{\mu_{U^{(j)}_{\mathbf{k}}}, \Lambda_{U^{(j)}_{\mathbf{k}}}\}$ for $1 \le j \le N$. [33] notes that one can avoid the difficulty underlying the estimation of these parameters through a fully Bayesian approach, complemented with a sampling-based Markov Chain Monte Carlo (MCMC) method to address the lack of the analytical solution.

7.5.2 *Noise-Sensitive Sample Assignment*

One crucial piece of information that the basic grid-based decomposition process fails to account for is *potentially available knowledge* about the distribution of the noise across the input tensor. As discussed earlier, a sub-tensor which is poorly decomposed due to noise may negatively impact decomposition accuracies also for other parts of the tensor. Consequently, it is important to allocate resources to prevent a few noisy sub-tensors from negatively impacting the overall accuracy.

We note that there is a direct relationship between the amount of noise a sub-tensor has and the number of Gibbs samples it requires for accurate decomposition. In fact, the numbers of Gibbs samples allocated to different sub-tensors $\mathcal{X}_{\mathbf{k}}$ in

Algorithm 2 *do not need to be the same.* As we have seen in Sect. 7.5.1, Phase 1 decomposition of each sub-tensor is independent from the others and, thus, the number of Gibbs samples of different sub-tensors can be different. In fact, more samples can provide better accuracy for noisy sub-tensors and this can be used to improve the overall decomposition accuracy for a given number of Gibbs samples. Consequently, given a set of sub-tensors, with different amounts of noise, uniform assignment of the number of samples, $L = \left(\frac{L_{(\text{total})}}{|\mathcal{K}|}\right)$, where $L_{(\text{total})}$ is the total number of samples for the whole tensor and $|\mathcal{K}|$ is the number of sub-tensors, may not be the best choice. In this chapter, we rely on this key observation to help assign Gibbs samples to the various sub-tensors. On the other hand, the number of samples also directly impacts the cost of the probabilistic decomposition process. Therefore, the sample assignment process must be regulated carefully.

7.5.2.1 Naive Option: Noise Density-Based Sample Assignment

Intuitively, the number of samples a noisy sub-tensor, $\mathcal{X}_{\mathbf{k}}$, is allocated should be proportional to the density, $nd_{\mathbf{k}}$, of noise it contains:

$$L(\mathcal{X}_{\mathbf{k}}) = \lceil \gamma \times nd_{\mathbf{k}} \rceil + L_{\min}, \tag{7.6}$$

where $L_{\min}$ is the minimum number of samples a (non-noisy) tensor of the given size would need for accurate decomposition and γ is a control parameter. Note that the value of γ is selected such that the total number of samples needed is equal to the number, $L_{(\text{total})}$, of samples allocated for the whole tensor:

$$L_{(\text{total})} = \sum_{\mathbf{k} \in \mathcal{K}} L(\mathcal{X}_{\mathbf{k}}). \tag{7.7}$$

7.5.2.2 SIG-Based Sample Assignment: S-Strategy

Equations (7.6) and (7.7), above, help allocate samples across sub-tensors based on their noise densities. However, as discussed earlier, inaccuracies in decomposition of one sub-tensor can propagate across the rest of the sub-tensors in Phase 2. Therefore, a better approach would be to consider how errors can propagate across sub-tensors when allocating samples. More specifically, if we could assign a significance score to each sub-tensor, $\mathcal{X}_{\mathbf{k}}$, that takes into account not only its noise density, but also the position of the sub-tensor relative to other sub-tensors, we could use this information to better allocate the Gibbs samples to sub-tensors.

As discussed earlier in Sect. 7.3.3, the sub-tensor impact graph (SIG) of a given tensor can be used for assigning impact scores to each sub-tensor. This process, however, requires (in addition to the given SIG) a seed node set, $S \subseteq V$, which serves as the context in which scores are assigned: Given the SIG graph, $G(V, E)$, and a set, $S \subseteq G(V, E)$, of seed nodes, the score $\mathbf{p}[i]$ of a node $v_i \in G(V, E)$ is obtained by solving $\mathbf{p} = (1 - \beta)\mathbf{A}\ \mathbf{p} + \beta\mathbf{s}$, where $\mathbf{A}$ denotes the transition matrix, β is a parameter controlling the overall importance of the seeds, and $\mathbf{s}$ is a seeding vector.

Our intuition is that we can use the sub-tensors with noise as the seeds in the above process. The naive way to create this seeding vector is to set $\mathbf{s}[i] = \frac{1}{\|S\|}$ if $v_i \in S$, and to $\mathbf{s}[i] = 0$, otherwise. However, we note that we can do better: given noise densities (nd) of the sub-tensors we can create a seeding vector

$$\mathbf{s}[\mathbf{k}] = \frac{nd_{\mathbf{k}}}{\sum_{\mathbf{j} \in \mathcal{K}} nd_{\mathbf{j}}},$$

and then, once the sub-tensor impact scores ($\mathbf{p}$) are computed, we can associate a noise-sensitive significance score,

$$\eta_{\mathbf{k}} = \frac{\mathbf{p}[\mathbf{k}] - min_{\mathbf{j} \in \mathcal{K}}(\mathbf{p}[\mathbf{j}])}{max_{\mathbf{j} \in \mathcal{K}}(\mathbf{p}[\mathbf{j}]) - min_{\mathbf{j} \in \mathcal{K}}(\mathbf{p}[\mathbf{j}])},$$

to each sub-tensor $\mathcal{X}_{\mathbf{k}}$. Given this score, we can then rewrite Eq. (7.6) as

$$L(\mathcal{X}_{\mathbf{k}}) = \lceil \gamma \times \eta_{\mathbf{k}} \rceil + L_{\min}. \tag{7.8}$$

7.5.3 *Evaluation*

In this section, we report experiments that aim to assess the proposed noise-sensitive sample assignment strategy (`s-strategy`) by comparing the performance of `nTD`, which leverages this strategy, against `GPTD` with uniform sample assignment and other naive strategies

7.5.3.1 Setup

Data Sets In these experiments, we used one user centered data set:*Ciao* [29]. The data is represented in the form of $5000 \times 5000 \times 996$ (density 1.7×10^{-6}) tensor. Its schema is $\langle user, item, time \rangle$. In this data, the tensor cells contain rating values between 1 and 5 or (if the rating does not exist) a special "null" symbol.

Noise In these experiments, uniform value-independent type of noise was introduced by modifying the existing ratings in the data set. More specifically, given a uniform noise profile and density, we have selected a subset of the existing ratings (ignoring "null" values) and altered the existing values—by selecting a completely new rating (which we refer to as *value-independent noise*).

Evaluation Criteria We use the *root mean squares error* (RMSE) inaccuracy measure to assess the decomposition effectiveness. We also report the decomposition times. Unless otherwise reported, the execution time of the overall process is reported as if sub-tensor decompositions in Phase 1 and Phase 2 are all executed serially, without leveraging any sub-tensor parallelism. Each experiment was run ten times with different random noise distributions and averages are reported.

Hardware and Software We used a quad-core CPU Nehalem Node with 12.00GB RAM. All codes were run using Matlab R2015b. For conventional CP decomposition, we used MATLAB Tensor Toolbox Version 2.6 [1].

7.5.3.2 Discussion of the Results

We start the discussion of the results by studying the impact of the `s-strategy` for leveraging noise profiles.

Impact of Leveraging Noise Profiles In Fig. 7.9, we compare the performance of `nTD` with noise-sensitive sample assignment (i.e., `s-strategy`) against `GPTD` with *uniform* sample assignment and the two naive noise adaptations, presented in Sects. 7.5.2.1 and 7.5.2.2, respectively. Note that in the scenario considered in this figure, we have 640 total Gibbs samples for 64 sub-tensors, providing on the average 10 samples per sub-tensor. In these experiments, we set $L_{\min}$ to 9 (i.e., very close to this average), thus requiring that $576 (= 64 \times 9)$ samples are uniformly distributed across the sub-tensors—this leaves only 64 samples to be distributed adaptively across the sub-tensors based on the noise profiles of the sub-tensors and their relationships to other sub-tensors. As we see in this figure, the proposed `nTD` is able to leverage these 64 uncommitted samples to significantly reduce RMSE relative to `GPTD` with *uniform* sample assignment. Moreover, we also see that naive noise adaptations can actually hurt the overall accuracy. `nTD`-naive uses biased sampling on the noise blocks and focuses on the centrality of sub-tensors. Thus, `nTD`-naive performs worse than uniform way. These together show that the proposed `s-strategy` is highly effective in leveraging rough knowledge about noise distributions to better allocate the Gibbs samples across the tensor.

In summary, the proposed sub-tensor impact graphs help allocate Gibbs samples in a way that takes into account how errors due to noise propagate across the whole tensor during the decomposition process.

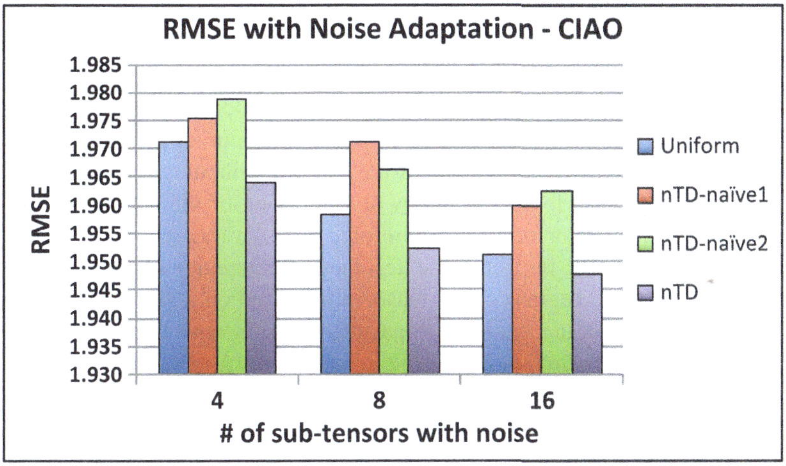

(a)

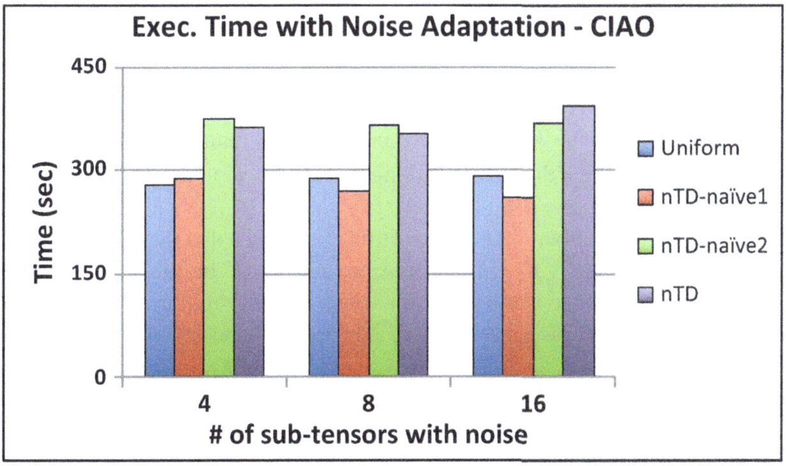

(b)

Fig. 7.9 RMSE and execution time (without sub-tensor parallelism) for nTD with different num. of noisy sub-tensors ($4 \times 4 \times 4$ grid; uniform noise; value independent noise; noise density 10%; total num. of samples = 640; $L_{\min} = 9$, $F = 10$; max. num. of P2 iteration = 1000). (**a**) RMSE for Ciao. (**b**) Time for Ciao

7.6 Application #3: Personalized Tensor Decomposition (PTD) and Rank Assignment Based on Sub-Tensor Impact Scores

In many clustering applications, the user may have a focus of interest, i.e., part of the data for which the user needs high accuracy, and beyond this area focus, accuracy may not be as critical. Relying on this observation, in this section, we present a personalized tensor decomposition (PTD) mechanism for accounting for the user's focus. Intuitively, the personalized tensor decomposition (PTD) algorithm partitions the tensor into multiple regions and then assigns different ranks to different sub-tensors: naturally, the higher the target rank is, the more accurate the decomposition of the sub-tensor. However, we note that preserving accuracy for foci of interest, while relaxing accuracy requirements for the rest of the input tensor is not a trivial task, especially because loss of accuracy at one region of the tensor may impact accuracies at other tensor regions. Therefore, PTD leverages sub-tensor impact graphs to account for the impact of the accuracy of one region of the tensor to the accuracies of the other regions of the tensor, each based on a different assumption about how the impact of inaccuracies propagates along the tensor. In particular, PTD analyzes the sub-tensor impact graph (in the light of the user's interest) to identify initial decomposition ranks for the sub-tensors in a way that will boost the final decomposition accuracies for partitions of interest.

7.6.1 Problem Formulation

Let an N-mode tensor $\mathcal{X} \in \mathbb{R}^{I_1 \times I_2 \times \ldots \times I_N}$ be partitioned into a set (or grid) of sub-tensors $\mathfrak{X} = \{\mathcal{X}_{\mathbf{k}} \mid \mathbf{k} \in \mathcal{K}\}$, where K_i indicates the number of partitions along mode-i, the size of the jth partition along mode i is $I_{j,i}$ (i.e., $\sum_{j=1}^{K_i} I_{j,i} = I_i$), and $\mathcal{K} = \{[k_{j_1}, \ldots, k_{j_i}, \ldots, k_{j_N}] \mid 1 \leq i \leq N,\ 1 \leq j_i \leq K_i\}$ is a set of sub-tensor indexes. The number, $\|\mathfrak{X}\|$, of partitions (and thus also the number, $\|\mathcal{K}\|$, of partition indexes) is $\prod_{i=1}^{N} K_i$. In addition, let $\mathcal{K}_P \subseteq \mathcal{K}$ be the set of sub-tensor indexes that indicate those sub-tensors for which the user requires higher accuracy. Note that, without loss of generality, we assume that the tensor $\mathcal{X}$ is re-ordered before it is partitioned in such a way that the number, K_i, of resulting partitions along each mode-i is minimal—i.e., along each mode, the entries of interest are clustered together to minimize the number of partitions.[6]

Given the above, let us use $\mathcal{X}_P$ as a shorthand to denote the cells of $\mathcal{X}$ collectively covered by the sub-tensors indexed by $\mathcal{K}_P$. The goal of the personalized tensor decomposition (PTD) is to obtain a *personalized (or preference sensitive) decomposition* $\hat{\mathcal{X}}$ of $\mathcal{X}$ in such a way that

$$accuracy(\mathcal{X}_P, \hat{\mathcal{X}}_P) > accuracy(\mathcal{X}_P, \tilde{\mathcal{X}}_P),$$

where $\hat{\mathcal{X}}_P$ is the reconstruction of the user selected region from the personalized decomposition $\hat{\mathcal{X}}$ and $\tilde{\mathcal{X}}_P$ is the reconstruction of the same region from a decomposition of $\mathcal{X}$ insensitive to the user preference $\mathcal{K}_P$ (or equivalently $\mathcal{K}_P = \mathcal{K}$). Naturally, we also aim that the time to obtain the personalized decomposition $\hat{\mathcal{X}}$ will be lesser than the time needed to obtain preference insensitive decomposition

[6]While this minimality criterion is not strictly required, the fewer partitions there are, the faster and potentially more effective will be the personalization process.

Algorithm 3 Overview of the PTD-CP process

Input: original tensor $\mathcal{X}$, partitioning pattern $\mathcal{K}$, user's focus $\mathcal{K}_P$, and decomposition rank, F
Output: personalized CP tensor decomposition $\hat{\mathcal{X}}$

1- compute the sub-tensor impact graph, $G(v, E, w())$, using the original tensor $\mathcal{X}$ and partitioning pattern $\mathcal{K}$;

2- for each partition $\mathbf{k} \in \mathcal{K}$, assign an initial decomposition rank $F_{\mathbf{k}}$ based on the sub-tensor impact graph, $G(v, E, w())$, and user's personalization focus $\mathcal{K}_P$;

3- obtain a block-based CP decomposition, $\hat{\mathcal{X}}$, of $\mathcal{X}$ using the partitioning pattern $\mathcal{K}$ and the initial decomposition ranks $\{F_{\mathbf{k}} \mid \mathbf{k} \in \mathcal{K}\}$;

return $\hat{\mathcal{X}}$

$\tilde{\mathcal{X}}$ and also that the personalized decomposition minimally impacts the rest of the tensor, i.e.,

$$accuracy(\mathcal{X}, \hat{\mathcal{X}}) \sim accuracy(\mathcal{X}, \tilde{\mathcal{X}}).$$

7.6.2 *Sub-Tensor Rank Flexibility*

Remember from Sect. 7.2.4, where we presented the update rules block-based tensor decomposition algorithms use for stitching the individual sub-tensor decompositions into a complete decomposition for the whole tensor, that (as visualized in Fig. 7.6) each $A^{(i)}_{(k_i)}$ is maintained incrementally by using, for all $1 \leq j \leq N$, the current estimates for $A^{(j)}_{(k_j)}$ and the decompositions in $\mathfrak{U}^{(j)}$, i.e., the F-rank subfactors of the sub-tensors in $\mathfrak{X}$ along the different modes of the tensor. A closer look at the update rule for $A^{(i)}_{(k_i)}$ further reveals the following observation:

- *Sub-tensor Rank Flexibility:* One critical observation is that the above formulation *does not require that all sub-tensors in $\mathfrak{X}$ are decomposed with the same target rank F.*

  In fact, as long as one sub-tensor is decomposed to rank F, all other sub-tensors can be decomposed to ranks lesser than F and we can still obtain full F-rank factors, $A^{(i)}$, for $\mathcal{X}$.

7.6.3 *Rank Assignment for Personalized Tensor Decomposition*

The PTD algorithm relies on this sub-tensor rank flexibility property to personalize the block-based (CP) decomposition process described earlier: unlike the basic block-based scheme discussed in the introduction, PTD improves the accuracy of the high-priority sub-tensors (indicated by $\mathcal{K}_P$) by assigning them higher initial decomposition ranks than the rest of the partitions (Algorithm 3). The **key difficulty**, however, is that one cannot arbitrarily reduce the decomposition ranks of low priority partitions, because the accuracy in one partition may impact final decomposition accuracies of other tensor partitions. As visualized in Algorithm 3, the proposed PTD algorithm first constructs a *sub-tensor impact graph*, G, that accounts for the propagation of inaccuracies along the tensor during a block-based decomposition process. The PTD algorithm then leverages this graph to account for the impact of the initial decomposition inaccuracy of one sub-tensor on the final decomposition accuracy of $\mathcal{X}_P$, i.e., the cells of $\mathcal{X}$ collectively covered by the user's declaration of interest (i.e., $\mathcal{K}_P$).

Intuitively, the initial decomposition rank, $F_{\mathbf{k}}$, of sub-tensor $\mathcal{X}_{\mathbf{k}}$ will need to reflect the impact of the initial decomposition of the sub-tensor $\mathcal{X}_{\mathbf{k}}$ on the final decomposition of the high-priority sub-tensors, $\mathcal{X}_{\kappa}$, $\kappa \in \mathcal{K}_P$. This implies that, when picking the decomposition ranks, $F_{\mathbf{k}}$, we need to measure how inaccuracies propagate within G over a large number of iterations of the alternating least squares (ALS) process. For this purpose we rely on the sub-tensor impact scores introduced in Sect. 7.3; more specifically, we compute the initial decomposition rank $F_{\mathbf{k}}$ of $\mathcal{X}_{\mathbf{k}}$ as

$$F_{\mathbf{k}} = \left\lceil F \times \frac{\mathbf{s}[k]}{\max_h\{\mathbf{s}[h]\}} \right\rceil,$$

where $\mathbf{s}[k]$ denotes the sub-tensor impact score of the sub-tensor $\mathcal{X}_{\mathbf{k}}$ in G. Intuitively, this formula sets the initial decomposition rank of the sub-tensor with the highest sub-tensor impact score (i.e., highest accuracy impact on the set of sub-tensors chosen by the user) to F, whereas other sub-tensors are assigned progressively smaller ranks (potentially all the way down to 1)[7] based on how far they are from the seed set in the sub-tensor impact graph, G.

7.6.4 Evaluation

In this section, we report sample results that aim to assess the effectiveness of the proposed personalized tensor decomposition (PTD) approach in helping preserve the tensor decomposition accuracy at parts of the tensor that are high-priority for the users.

[7]It is trivial to modify this equation such that the smallest rank will correspond to a user provided lower bound, $F_{\min}$, when such a lower bound is provided by the user.

Table 7.1 Various tensor partitioning scenarios considered in the experiments: the percentages are the sizes of the partitions (relative to the overall size of the mode) along each mode

2 × 2 × 2 partitions		
Configuration	Part. #1	Part. #2
1 - Most balanced	50%	50%
2 -	40%	60%
3 -	30%	70%
4 -	25%	75%
5 -	10%	90%
6 - Least balanced	1%	99%

7.6.4.1 Setup

Data Set In the experiments reported in this chapter, we used the *Ciao* [29] data set, represented in the form of a $167 \times 967 \times 18$ (density 2.2×10^{-4}) tensor, with the schema $\langle user, item, category \rangle$. In these experiments, we assume that the input tensor is partitioned into 8 ($= 2 \times 2 \times 2$) according to the scenarios shown in Table 7.1 and two randomly selected sub-tensors are marked as more important than the rest.

Decomposition Strategies We considered five decomposition strategies: not personalized (`NP`), uniform edge weights (`UNI`), surface of interaction-based edge weights (`SURF`), value alignment-based edge weights (`VAL`), and combined edge weights (`COMB`). The target CP decomposition rank, F, is set to 10.

Evaluation Criteria We use the measure reported in Sect. 7.2.2.2 to assess decomposition accuracy and execution time. In particular, we report accuracies for both user's area of focus and the whole tensor.

Hardware and Software We ran the experiments reported in this section on a quad-core Intel(R) Core(TM)i5-2400 CPU @ 3.10GHz machine with 8.00GB RAM. All codes were implemented in Matlab and run using Matlab R2015b. For CP decomposition, we used MATLAB Tensor Toolbox Version 2.6 [1].

7.6.4.2 Discussion of the Results

Sub-tensor impacts help take into account how inaccuracies in the decomposition propagate into high- and-low-priority regions. We therefore expect that allocating resources using sub-tensor impact graphs should provide better accuracies for high-priority regions. As we see in Fig. 7.10, as expected, PTD algorithms boost accuracy for the high-priority partitions in the user focus, especially where the partitions are of heterogeneous sizes (as is likely to be the case in real situations). While, as would be expected, the PTD algorithms have impact on the overall decomposition accuracy for the whole tensor, this is more than compensated by gains in accuracies in high-priority areas. Moreover, the figure also shows that the gains in accuracy in high-priority partitions within the user's focus come also with significant gains in execution times for the decomposition process.

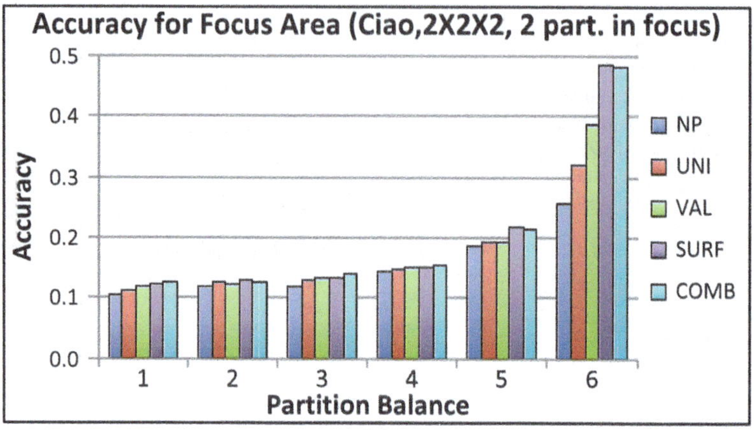

(a)

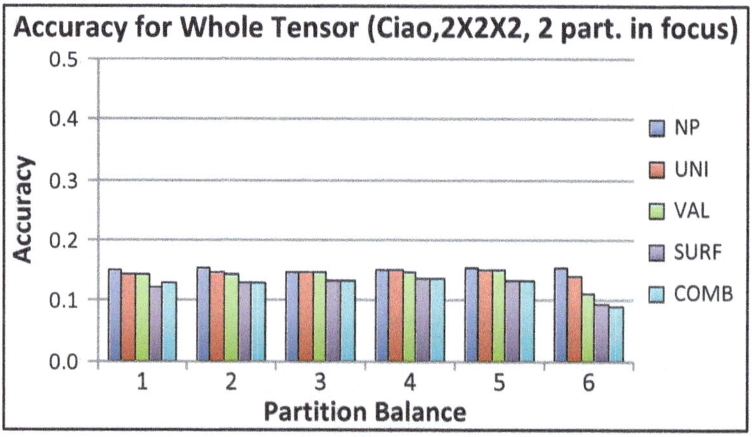

(b)

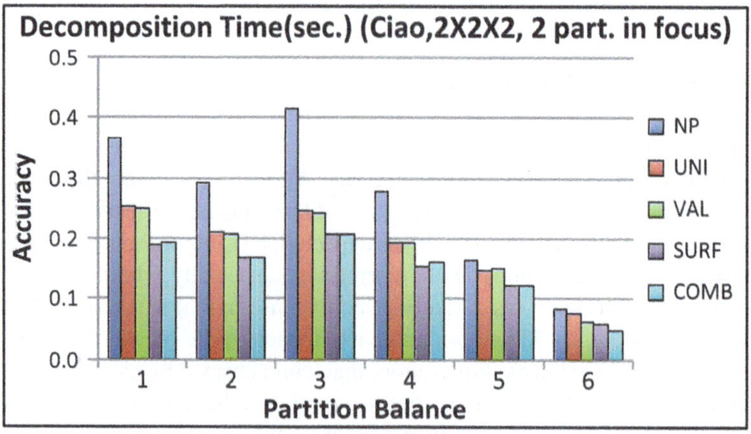

(c)

Fig. 7.10 Experiment results with two partitions in focus. (**a**) Accuracy of the region of focus. (**b**) Whole tensor accuracy. (**c**) Decomposition time

The figure also establishes that, both in terms of accuracy and execution time gains, we can order the various edge weighting strategies as follows: `UNI` (least effective), `VAL`, `SURF`, and `COMB` (most effective). In other words, as we argued in Sect. 7.3.2.4, the most effective way to account for the propagation of inaccuracies is to combine the surface of interaction and value alignment-based edge weights into a single weight which accounts for both shapes of the sub-tensors and their value alignments.

7.7 Conclusions

Computational complexity of tensor decomposition is a major bottleneck in many applications. Block-based tensor decomposition is employed to efficiently conduct tensor decomposition for large-scale data analysis. However, we need a smart strategy to account for the relationship among these sub-tensors(blocks). Therefore, we proposed sub-tensor impact graph (SIG) to account for the propagation of impacts within the sub-tensors during the decomposition process. Then, we presented three applications of SIG to efficiently solve the challenges in personalized tensor analysis, incremental tensor analysis, and noise tensor analysis. Experiments results on real data sets show that SIGs can improve the performance of tensor analysis in these three applications in both of execution time and decomposition accuracy.

Finally, we would like to note that, here we presented three distinct uses of sub-tensor impact graphs for three distinct challenges (dynamic data, noisy data, and personalization). In practice, there is no reason these approaches cannot be combined to tackle more complex scenarios, such as personalized clustering and analysis over dynamically evolving data sets. We leave the study of these more complex scenarios as future work.

Acknowledgements Research is supported by NSF#1318788 "Data Management for Real-Time Data Driven Epidemic Spread Simulations," NSF#1339835 "E-SDMS: Energy Simulation Data Management System Software," NSF#1610282 "DataStorm: A Data Enabled System for End-to-End Disaster Planning and Response," NSF#1633381 "BIGDATA: Discovering Context-Sensitive Impact in Complex Systems," and "FourCmodeling": EUH2020 Marie Sklodowska-Curie grant agreement No 690817.

References

1. B.W. Bader, T.G. Kolda et al., MATLAB Tensor Toolbox Version 2.5. Available online (January 2012)
2. A. Balmin, V. Hristidis, Y. Papakonstantinou. ObjectRank: authority-based keyword search in databases, in *Proceedings of the 30th International Conference on Very Large Data Bases (VLDB)* (2004)

3. X. Cao, X. Wei, Y. Han, D. Lin, Robust face clustering via tensor decomposition. IEEE Trans. Cybern. **45**(11), 2546–2557 (2015)
4. S. Chakrabarti, Dynamic personalized pagerank in entity-relation graphs, in *Proceeding WWW '07 Proceedings of the 16th International Conference on World Wide Web* (2007)
5. X. Chen, K.S. Candan, LWI-SVD: low-rank, windowed, incremental singular value decompositions on time-evolving data sets, in *KDD '14 Proceedings of the 20th ACM SIGKDD International Conference on Knowledge Discovery and Data Mining* (2014)
6. I. Davidson, S. Gilpin, O. Carmichael, P. Walker, Network discovery via constrained tensor analysis of FMRI data, in *19th ACM SIGKDD International Conference on Knowledge Discovery and Data Mining*, pp. 194–202 (2013)
7. C. Ding, X. He, K-means clustering via principal component analysis, in *ICML '04 Proceedings of the Twenty-First International Conference on Machine Learning* (2004)
8. P. Drineas, A. Frieze, R. Kannan, S. Vempala, V. Vinay, Clustering large graphs via the singular value decomposition. Mach. Learn. **56**, 9–33 (2004)
9. F.M. Harper, J.A. Konstan, The MovieLens datasets: history and context. Trans. Interact. Intell. Syst. **5**, 19:1–19:19 (2015)
10. R.A. Harshman, Foundations of the PARAFAC procedure: model and conditions for an explanatory multi-mode factor analysis. *UCLA Working Papers in Phonetics*, vol. 16 (1970), pp. 1–84
11. S. Huang, K.S. Candan, M.L. Sapino, BICP: block-incremental CP decomposition with update sensitive refinement, in *Proceeding CIKM '16 Proceedings of the 25th ACM International on Conference on Information and Knowledge Management* (2016)
12. I. Jeon, E. Papalexakis, U. Kang, C. Faloutsos, HaTen2: billionscale tensor decompositions, in *Proceedings - International Conference on Data Engineering (ICDE)* (2015)
13. B. Jeon, I. Jeon, L. Sael, U. Kang, SCouT: scalable coupled matrix-tensor factorization - algorithm and discoveries, in *IEEE 32nd International Conference on Data Engineering (ICDE)* (2016)
14. U. Kang, E.E. Papalexakis, A. Harpale, C. Faloutsos, Gigatensor: scaling tensor analysis up by 100 times algorithms and discoveries, in *Proceedings of the ACM SIGKDD International Conference on Knowledge Discovery and Data Mining (KDD)*, pp. 316–324 (2012)
15. M. Kim, K.S. Candan, Decomposition by normalization (DBN): leveraging approximate functional dependencies for efficient CP and tucker decompositions. Data Min. Knowl. Disc. **30**(1), 1–46 (2016)
16. T.G. Kolda, B.W. Bader, Tensor decompositions and applications. SIAM Rev. **51**(3), 455–500 (2009)
17. T.G. Kolda, J. Sun, Scalable tensor decompositions for multi-aspect data mining, in *Eighth IEEE International Conference on Data Mining (ICDM)* (2008)
18. X. Li, S.Y. Huang, K.S. Candan, M.L. Sapino, Focusing decomposition accuracy by personalizing tensor decomposition (PTD), in *Proceeding CIKM '14 Proceedings of the 23rd ACM International Conference on Conference on Information and Knowledge Management* (2014)
19. X. Li, K.S. Candan, M.L. Sapino, nTD: noise-profile adaptive tensor decomposition, in *Proceeding WWW '17 Proceedings of the 26th International Conference on World Wide Web* (2017)
20. S. Papadimitriou, J. Sun, C. Faloutsos, Streaming pattern discovery in multiple time-series, in *Proceeding VLDB '05 Proceedings of the 31st International Conference on Very Large Data Bases* (2015)
21. E. Papalexakis, C. Faloutsos, N. Sidiropoulos, Parcube: sparse parallelizable tensor decompositions, in *Joint European Conference on Machine Learning and Knowledge Discovery in Databases (ECML PKDD)*, pp. 521–536 (2012)
22. I. Perros, E.E. Papalexakis, F. Wang, R. Vuduc, E. Searles, M. Thompson, J. Sun, Spartan: scalable parafac2 for large & sparse data (2017). arXiv preprint arXiv:1703.04219
23. A.H. Phan, A. Cichocki, PARAFAC algorithms for large-scale problems. Neurocomputing **74**(11), 1970–1984 (2011)
24. C.E. Priebe et al., Enron data set (2006). http://cis.jhu.edu/parky/Enron/enron.html

25. R. Salakhutdinov, A. Mnih, Probabilistic matrix factorization, in *Proceeding NIPS'07 Proceedings of the 20th International Conference on Neural Information Processing Systems* (2007)
26. J. Sun, S. Papadimitriou, P.S. Yu, Window based tensor analysis on high dimensional and multi aspect streams, in *Sixth International Conference on Data Mining (ICDM'06)*, pp. 1076–1080 (2006)
27. J. Sun, D. Tao, S. Papadimitriou, P.S. Yu, C. Faloutsos, Incremental tensor analysis: theory and applications. ACM Trans. Knowl. Discov. Data **2**(3), Article No. 11 (2008)
28. Y. Sun, J. Gao, X. Hong, B. Mishra, B. Yin, Heterogeneous tensor decomposition for clustering via manifold optimization. IEEE Trans. Pattern Anal. Mach. Intell. **38**(3), 476–489 (2016)
29. J. Tang et al., Trust & distrust computing dataset (2011). https://www.cse.msu.edu/~tangjili/trust.html
30. C.E. Tsourakakis, Mach: fast randomized tensor decompositions (2009). Arxiv preprint arXiv:0909.4969
31. L. Tucker, Some mathematical notes on three-mode factor analysis. Psychometrika **31**, 279–311 (1966)
32. J. Wu, Z. Wang, Y. Wu, L. Liu, S. Deng, H. Huang, Tensor CP decomposition method for clustering heterogeneous information networks via stochastic gradient descent algorithms. Sci. Program. **2017**, 13 (2017), Article ID 2803091
33. L. Xiong et al., Temporal collaborative filtering with Bayesian probabilistic tensor factorization, in *Proceedings of the 2010 SIAM International Conference on Data Mining* (2010)

Index

O. Nasraoui, C.-E. Ben N'Cir (eds.), *Clustering Methods for Big Data Analytics*, Unsupervised and Semi-Supervised Learning,
https://doi.org/10.1007/978-3-319-97864-2

The manufacturer's authorised representative in the EU is Springer Nature Customer Service Centre GmbH, Europaplatz 3, 69115 Heidelberg, Germany. If you have any concerns regarding our products, please contact ProductSafety@springernature.com

Printed and bound by CPI Group (UK) Ltd, Croydon, CR0 4YY
30/09/2026
02225230-0001